THE SEED COSMOLOGY

Revelations
of the
Seed Origin
of the
Universe

BY

KAZMER UJVAROSY

CONTENTS

… civilization is founded on some sort of theory of the universe, and can be restored only through a spiritual awakening and a will for ethical good in the mass of mankind …

Albert Schweitzer, *The Decay and the Restoration of Civilization* (London: A. & C. Black, 1923)

What is the worst thing the Evil Urge can achieve? To make man forget that he is the son of a king.

Rabbi Shlomo Halevi of Karlin (1738 – 1792).

If the sow with her snout should happen to imprint the letter A upon the ground, wouldst thou therefore imagine that she could write out a whole tragedy as one letter?

Sir Francis Bacon (1561 – 1626)
Interpretation of Nature

The greatest science fiction stories are in the science of physics.

Morris Kline, *Mathematics and the Search for Knowledge* (New York: Oxford University Press, 1985)

PREFACE

In common religious thought God is conceived to be the creator, all-powerful and all-wise governor of the universe, the source of all created things, the incorporeal creative essence that rules over all as eternal spirit. The atheists, in opposition to the religious view, deny the reality of an everlasting world spirit or creative agent. In their opinion God does not exist because matter is the only reality, and if matter is the only reality, then an incorporeal being cannot be real.

The primary purpose of this treatise is to eliminate a common and central misconception from the religious and atheist world views. This misconception is the belief that God is incorporeal. An examination of the sacred and secret traditions of the world reveals that when the religious authorities speak of God, essentially they do not refer to an incorporeal creative principle, but rather to that formative and controlling hereditary substance which today we call genetic material.

The traditions of the world indicate, in other words, that the observation that a tree or biological system is the seed's way of making seeds in its own image or likeness may have led to the realization that similarly the world system is a seed's way of making human beings in its own image or likeness. The inferred seed, Creator or God of the universe is often referred to by such terms as "cosmic seed," "holy seed," "procreative seed of the world," "germ of the cosmos," or "immortal seed."

The finding that the Creator of the universe may be genetic in nature is significant, it implies that the notion of a Creator of the universe is a scientific concept, i.e. falls within the realm of empirical science.

It seems evident that the fundamental difference between the ancient cosmologies and the current origin models is that whereas religion credits the creation of the universe to the seed of the universe or existing highest form of life, the evolutionary Big Bang cosmology credits the origin of the universe to the assumed explosion of a super-dense, inanimate, and dimensionless singularity.

These two basic origin models of the universe indicate that science needs the insight provided by religion regarding the seed of the universe, and religion needs the insight provided by science regarding the cosmic system's procreative functions. It may be concluded that neither religion, nor science, is complete without the other.

Hopefully the information contained in this paper will help to bridge the gulf separating religion and science, and will open the way for new discoveries.

Chapter One

BIOLOGICAL SYSTEMS: MODELS FOR THE COSMOS

The English novelist Samuel Butler solved the riddle about the priority of the chicken or the egg with the neat answer: "The hen is the egg's way of making another egg." Thus in Butler's view the seeds or germ cells of biological systems exist only to perpetuate themselves. The fertilized egg produces a biological system in the form of the hen, and through that system reproduces itself in the form of another egg. This reproduction of the original or parent egg may be called the genetic output or end product of the hen's biological system.

The same process is observable in the case of trees. The parent seed produces a tree, and through that biological system reproduces itself in the form of other seeds. In general these reproductions of the parent seed are contained in the fruit—i.e., in the tree's end product—, but only a single seed is needed to produce a tree many billions of times its own size and weight.

The lesson of this reasoning is that only the fruit, the tree's end product, contains in itself the genetic information or qualities required for the tree's production. The tree's fruit or end product, in other words, contains the reproductions of that parent seed that created the tree. Or, stated differently, the seed of a tree is both the producer and the product of the tree.

Peter Sterry, a leading Cambridge Platonist, put it this way: "All *Philosophy* agreeth in this, that the *last end* is the *first mover*." This principle, when applied to the universe and man, suggests that if man is the highest product, the end product, or "last end" of the universe—and there is no evidence to the contrary—then his seed is the cause or "first mover" of the universe. And just as the seed in the fruit is both the producer and the product of the tree, so is man's seed or genome both the producer and the product of the universe.

In this analogy the tree represents the universe; the tree's fruit represents man—i.e., the end product of the universe; and the seed in the fruit represents man's seed or genetic constitution.

The point of this analogy is that by comparing the tree—whose origin is already known—to the universe—whose origin is unknown—it is possible to suggest a probable explanation or theory for the origin of the universe.

Expressed somewhat differently, by knowing that the fruit of a tree contains a tiny seed that has the potential to produce an immensely greater tree, it is conceivable that man—if he is indeed the "crown" of creation—contains a seed of the universe that has the potential to bring a world into being.

To continue this reasoning, by knowing that all parts of a tree exist for the benefit of its fruit, it is conceivable that all parts of the universe exist for the benefit of man. From this point of view, then, man indeed may be regarded as the center of the universe.

The purpose of this paper is not only to propose that the above analogical reasoning may hold the key to the mystery of the origin of the universe, but also to show that many traditions and sacred writings explain the origin and nature of the universe in terms of an eternal seed; compare the universe to a Tree of Life; and claim that man, being the end product of the universe, contains in his body an image of that perpetual parent seed which created the universe for the purpose of self-reproduction.

The tree is made manifest by its fruit …

The Epistle of Ignatius to the Ephesians, III. 16

Chapter Two

CREATION FROM SEED

Perhaps the best way to demonstrate that in many world views the creator of the universe is conceived of as a cosmic seed is to present a number of creation stories from diverse societies in which an everlasting seed is central to the formation of the universe.

In a Sumerian myth the act of creation is credited to Enki, the great fresh-water god, who "fertilised the swampy land with his own seed." In a Polynesian tradition Taaroa plays the central role in creation. He is identified with an immortal seed.

> Alone existing, he changes himself into universe. Taaroa is the light, he is the seed, he is the base, he is the incorruptible. The universe is only the shell of Taaroa. It is he who puts it in motion and brings forth its harmony.[1]

In the world view of the Desana, an isolated Indian tribe of the Northwest Amazon, the birth of the universe is attributed to the fertilizing power of the Sun, "which has the characteristics of semen," comments Gerardo Reichel-Dolmatoff in his *Amazonian Cosmos*.[2]

> The Creation of the Universe was the result of the "yellow intention" or "yellow purpose" of the Sun. The yellow color, as we said, symbolizes semen among the Desana and plays an important role in the image that is held of the Universe, especially on ritual occasions.[3]

In *Symbols and Values in Zoroastrianism* Jacques Duchesne-Guillemin writes that in order to create all creatures Ahura Mazda possesses xvarenah, which is "both fiery and seminal" in nature, and that this procreative power or "life-giving fire is repeatedly stated to be 'in the seed'." He adds: "The xvarr of God is 'simply' his seed.[4]

In Chinese tradition the ultimate cause of the world is Tao. "Before heaven and earth were, Tao was," teaches the leading Taoist Chuang-tse. "It has existed without change from all time." In the Chinese meditation text *Hui Ming Ching* Tao is called "the primordial seed."[5]

---

[1] Quoted in Whitall N. Perry, *A Treasury of Traditional Wisdom* (New York: Simon and Schuster, 1971), p. 26.

[2] Gerardo Reichel-Dolmatoff, *Amazonian Cosmos: The Sexual and Religious Symbolism of the Tukano Indians* (Chicago University Press, 1971), p. 98.

[3] Ibid., p. 47.

[4] Jacques Duchesne-Guillemin, *Symbols and Values in Zoroastrianism* (New York: Harper & Row, 1966), pp. 142 and 145.

[5] The *Hui Ming Ching*, in Richard Wilhelm, trans. and expl., *The Secret of the Golden Flower: A Chinese Book of Life*, Carl G. Jung, forew. and comm. (New York: Harvest Book, 1962), p. 72.

In *The World-Conception of the Chinese* Alfred Forke writes that when it comes to creation, heaven and earth on the cosmic level act like husband and wife on the human level. He notes: "Heaven but gives the impulse of life, the earth receives his seed and in her bowels develops the young creature until its birth."[6]

In addition to these traditions the Zunis of New Mexico believe that the Sun-father, "the Maker and Container of All," existed before the beginning of time, and "formed the seed-stuff of twain worlds, impregnating therewith the great waters." In the Scandinavian story of the Universal Egg it is "the dark cleft in space in which the seed of the world is sown." In the mystic Greek religion of Orpheus the symbol for the active agent of creation is a serpent. It is depicted entwined about the passive agent, the orphic egg. "The serpent is the sperm," explains Manly P. Hall in his book, *Man: The Grand Symbol of the Mysteries*, "and the egg the ovum."[7]

> From the union of Ether and Chaos - -the germ and the egg - -the world comes into being, and by recourse to the cosmogony myths, which for the most part are founded upon embryology, we can secure a very adequate account of the development of the "world animal."
> From an occult standpoint, then, the spermatozoon is the carrier of the archetype. It is a little ark in which the seeds of life are carried upon the surface of the waters that at the appointed time they may replenish the earth. A triad of forces--spiritual, psychical, and material--are contained within the head of the sperm.[8]

The Yorubas of Nigeria have a creation story which identifies one of the deities with the human seed in the maternal womb. It says in part:

> At the beginning everything was water. Then Olodumare, the supreme god, sent Obatala (or Orisbanla) down from heaven, to create the dry land.... Obatala made man out of earth. After shaping men and women he gave them to Olodumare to blow in the breath of life.... Obatala is still the one who gives shape to the new babe in the mother's womb.[9]

In *The Dawn and Twilight of Zoroastrianism* R. C. Zaehner writes that prior to creation Ohrmazd, the creator, remained "in a moist state like semen," but by the act of creation this moist state mixed "like semen and blood." Zaehner adds: "Even now on earth do men in this wise grow together in their mother's womb, and are born and bred."[10]

---

[6] Alfred Forke, *The World-Conception of the Chinese* (Arno Press, 1975), p. 69.

[7] Manly P. Hall, *Man: The Grand Symbol of the Mysteries* (Los Angeles: The Philosophical Research Society, Inc., 1972), p. 76.

[8] Ibid., p. 84.

[9] Ulli Beier, *The Origin of Life and Death* (London, Ibadan, Nairobi: Heinemann, 1969), pp. 47-48.

[10] R. C. Zaehner, *The Dawn and Twilight of Zoroastrianism* (New York: G. P. Putnam's Sons, 1961), p. 250.

Ernest Busenbark concludes: "Endowed with inherent productive activity, it is the 'seminal reason' of the world which manifests itself in all phenomena of nature."[11]

Creation from Seed in Hinduism

The Hindu sacred writings also speak of the creation of the universe from a seed. In the *Mundaka Upanishad* the universe is derived from "the undecaying Seed"; in the *Rig Veda* "the primal seed" acted on the "unillumined water" which was "void and formless"; in the *Manu Samhita* also a seed transmitted into the primeval water brings about the creation; and in the *Atharvaveda* "Prana begets the universe." "Prana," writes Swami Nikhilananda, "... is the seed of all the tangible objects in the universe."[12]

Another Hindu scripture, the *Brihadaranyaka Upanishad*, provides the following account of creation:

> The world existed first as seed, which as it grew and developed took on names and forms. As a razor in its case or as fire in wood, so dwells the Self, the Lord of the universe, in all forms, even to the tips of the fingers. Yet the ignorant do not know Him, for behind the names and forms he remains hidden.[13]

The Hindu *Svetasvatara Upanishad* indirectly identifies the creator of the world with the human genome. It says: "He indeed, the Lord, who pervades all regions, was the first to be born, and it is He who dwells in the womb." Benjamin Walker notes:

> In Vajrayana and other forms of Tantric Buddhism, semen is equated with the highest bliss, and Buddha is represented as 'dwelling in the vagina of the female in the name of semen'. Buddha as semen 'represents neither existence nor non-existence, but the formless nature of Supreme Bliss'.[14]

In the story of the golden egg the self-existent Brahman, "the essence of all beings," writes Pierre Grimal, editor of the *Larousse World Mythology*, "put a seed" in the primordial waters on the eve of creation.[15] As a result of that creative act Brahma was born, the image of the Absolute or Universal Spirit.

---

[11] Ernest Busenbark, *Symbols Sex, and the Stars* (New York: Truth Seeker Co., 1949), p. 316.

[12] Swami Nikhilananda, *The Upanishads* (New York: Harper & Row, 1964), p. 78.

[13] Swami Prabhavananda and Frederick Manchester, selected and trans., *The Upanishads: Breath of the Eternal* (New York: Mentor Book, 1957), p. 80.

[14] Benjamin Walker, *Hindu World: An Encyclopedic Survey of Hinduism*, Vol. I (London: George All & Unwin Ltd., 1968), p. 153.

[15] Pierre Grimal, ed., *Larousse World Mythology*, 2nd ed. (London: Paul Hamlyn, 1969), p. 210.

Brahma is at the same time the hidden God in man, called *Atman* in Sanskrit. This *Atman*, Self, or God within, explains the *Aitareya Upanishad*, is "in man as a germ, which is called seed."[16]

If the reproduction of Brahman, "the Ultimate Cosmic Force," is present in man as the "Atman" or "Self," which is a germ or seed in man, then from this it may be inferred that man's genome or genetic constitution is God, the creator of the universe.

Creation from Seed in Judaism

With regard to the Holy Spirit—that according to the Book of Genesis moved over the primordial "waters" in the beginning, over what St. Augustine calls "formless matter"—Carl G. Jung notes that in the alchemical texts it is described as a "procreator" that impregnated "the waters with the seed of life," and also as that reproductive agent which fertilized the Virgin Mary.[17] Out of this follows that if the primeval waters were impregnated with "the seed of life," then Mary was impregnated with "the seed of life" as well. The result of the first fertilizing act by the Holy Spirit was the birth of the universe, and that of the second the birth of Jesus Christ, the progeny of the Holy Spirit.

It also may be concluded that the expression "Holy Spirit" is actually an ancient term for that hereditary substance which today is called germ cell, spermatozoon, or genetic material. As a matter of fact Jonn Mumford in *Sexual Occultism* writes that A. E. Waite in the *Holy Kabbalah* "gives a Latin quotation which is a most explicit statement that the Holy Spirit (Shekinah) exists within the genital parts."[18] In *Acupuncture* Felix Mann also notes that under the term "Spirit" the ancients understood the united human germ cells or zygote. According to the sage Ling Shu, Mann writes, "The origin of Life is in the Life Essence (the male and female semen). When these two unite to make one, that is called the Spirit."[19] In *Aurora Consurgens* Marie-Louise von Franz notes in connection with a different text:

> Similar views are reflected in a Hermetic fragment: from the combined breath of the opposed qualities arises a pneuma and sperma that corresponds to the divine pneuma. "From this creative pneuma the child is formed in the matrix."[20]

---

[16] Lewis Browne, *The World's Great Scriptures* (New York: The Macmillan Co., 1946), p. 67.
[17] Carl G. Jung, *Alchemical Studies*, R. F. C. Hull, trans., 2nd ed. (Princeton, NJ: Princeton University Press, 1970), p. 214.
[18] Jonn Mumford, *Sexual Occultism* (Saint Paul, MN: Llewellyn Publications, 1975), p. 9.
[19] Felix Mann, *Acupuncture* (New York: Vintage Books, 1973), p. 54.
[20] Marie-Louise von Franz, ed., *Aurora Consurgens: Commentary*, R. F. C. Hull and A. S. B. Glover, trans. (New York: Pantheon Books, 1966), p. 292.

In *A Physical Interpretation of the Universe: The Doctrines of Zeno the Stoic*, H. A. K. Hunt quotes a comment of Eusebius: "The sperm which man releases Zeno says to be *pneuma* with moisture."[21] Eduard Zeller in *Outlines of the History of Greek Philosophy* identifies this *pneuma* in the sperm—which is evidently the genetic information—with God, with that "corporeal" and "final world-cause" which "animates and moves all things." "Since everything in the world owes its qualities, its motion and its life to it," he writes, "it [the *pneuma*] must bear the same relation to the universe as soul to our body.[22]

If Zeller's interpretation is correct and the *pneuma* or final world-cause is related to the universe as soul to the human body, then the creator of the world must have a genetic nature, because the soul is derived from man's seed. "It would seem that the intellectual soul is produced from the semen," reasons St. Thomas Aquinas. "For it is written (Gen. XLVI. 26): All the souls that came out of Jacob's thigh, sixty-six. But nothing is produced from the thigh of a man, except from the semen."

Aristotle likewise says that the semen not only has soul, but is soul potentially. Here it may be added that according to Emanuel Swedenborg "a man has his beginning from the soul, which is the very essence of the seed."[23]

Today it is a well-known fact that the very essence of the seed is its genetic information. Consequently the soul must be synonymous with the human genome. Indeed, Géza Roheim in his *Australian Totemism* observes that "the spirit is a symbolical representative of the spermatozoon."[24] From this it follows that when Clement says that Jesus Christ, "being first a spirit, was made flesh," it really means that Jesus Christ, being first human genome, was made flesh.

It must be pointed out here that G. F. Nicolai in *The Biology of War* identifies the "sacred pneuma" of the Bible with the "germ-plasm" in the human body. He writes:

> Luther translated this [pneuma] by "der Geist, der lebendig macht" ("The spirit that quickeneth"), thus attributing a purely symbolical meaning to it. The conception of pneuma, however, goes beyond this, and cannot be understood save by those acquainted with its origin in Greek philosophy. Into this I am unable to enter in detail, but Diogenes Laertes expressly states, "That which causes the procreation of us all is the pneuma," thus meaning precisely what we may now call germ-plasm. Moreover, just as we must now make up our minds that an almost imponderable quantity of germ-plasm influences the whole body, even so the men of old imagined the mysterious workings of the "Holy Spirit."

---

[21] H. A. K. Hunt, *A Physical Interpretation of the Universe: The Doctrines of Zeno the Stoic* (Melbourne University Press, 1976), p. 52.

[22] Eduard Zeller, *Outlines of the History of Greek Philosophy*, 13th ed. (New York: Humanities Press, 1951), p. 216.

[23] Emanuel Swedenborg, quoted in D. Gopaul Chetty, *New Light on Indian Philosophy* (New York: E. P. Dutton & Co., MCMXXIII), p. 200.

[24] Géza Roheim, *Australian Totemism: A Psycho-Analytic Study in Anthropology*, M. D. Eder, intr. (London: G. Allen & Unwin Ltd., 1925), p. 167.

In the sixth chapter of the Gospel of St. John, verse 63, we read, "It is the spirit [pneuma] that quickeneth; the flesh profiteth nothing." Thus the Bible also must really be referring to germ plasma. Now, there is no need to state that this pneuma is never clearly expressed either in Greek writings or in the Bible what we now mean by germ plasm. Nevertheless, it is important to recollect that those who wrote the Bible felt, as it were, intuitively, that it existed.[25]

The equation of the maker of the universe with the human genome is supported by other data. For example John M. Allegro, a linguist and scholar on the Dead Sea Scrolls, derives the name of the God of the Hebrews, Yahweh or Jehovah, from a Sumerian term that means semen. In *The Sacred Mushroom and the Cross* he writes: "Thus the principal gods of the Greeks and Hebrews, Zeus and Yahweh (Jehovah), have names derived from Sumerian meaning 'juice of fecundity', spermatozoa, 'seed of life'." Again he notes: "The self-identification, 'I am Yahweh your God' merely states in normal Hebrew the Sumerian original of those divine names, E-LA-UIA, 'juice of fecundity; sperm'."[26]

This interpretation is supported by a statement in the mystical writing *Source of Measures*, where Jehovah identifies himself to Moses: "The summation of my name is *Sacr*, the carrier of the germ."

Much more data from various sources support the theory that the Bible's Holy Spirit or Spirit of God equals the human genome or essence of the human reproductive cells.

Now only one more information will be examined. It was provided by Avicenna, the Persian physician and philosopher. He identified "the spirit of the Lord which fills the whole world and in the beginning swam upon the waters" with Mercurius of the alchemists. "Mercurius is often called ... spiritus seminalis,"[27] remarks Jung, and he also notes: "Mercurius ... is the essence or 'seminal matter' of both man and woman." "He is the primordial matter from which God created all material things."[28]

Creation from Seed in Egyptian Mythology

In the Heliopolitan story of creation the ultimate cause of the world is Atum-Re (or Ra). This alleged sun-god and inexhaustible source of biospermatic energy describes his self-fertilizing act as follows:

---

[25] G. F. Nicolai, *The Biology of War* (New York: The Century Co., 1918), pp. 451-452.

[26] John M. Allegro, *The Sacred Mushroom and the Cross*, 2nd pr. (New York: Bantam Books, 1971), pp. 20 and 199.

[27] Jung, op. cit. note 17, p. 213.

[28] Carl G. Jung, *Mysterium Coniunctionis*, R. F. C. Hull, trans., 2nd ed. (Princeton, NJ: Princeton University Press, 1970), pp. 462 and 502.

I, even I, had union with my clenched hand, I joined myself in an embrace with my shadow, I poured seed into my mouth, my own, I sent forth issue in the form of Shu, I sent forth moisture in the form of Tefnut.[29]

This first pair of the gods, Shu and Tefnut, in turn gave birth to the Osiride family and all others. "But the first piece of solid matter actually created by Atum in the primeval ocean," remarks Henry Frankfort in his book, *Kingship and the Gods*, "... was a stone, the Benben; and it had originated from a drop of the seed of Atum which fell into the primeval ocean."[30]

According to an Egyptian treatise on magic Atum-Re had many names, but he had a secret and sacred one which made him all-powerful. The goddess Isis, the wife of Osiris, conjured Re to reveal to her his mysterious name, which eventually he did in the company of gods. He said:

Come to me, 0 my children, offspring of my body. I am a prince, the son of a prince, the divine seed of a god. My father devised my name; my father and my mother gave me my name, and it remained hidden in my body since my birth, that no magician might have magic power over me.[31]

These words make it fairly clear that Re's secret name is actually the "divine seed" he had received from his parents before his birth, and which "remained hidden" in his body. When in the *Papyrus of Nu* the god says, "'He that never suffereth corruption' is my name," he again alludes to his seminal nature.

Creation from Seed in Greek Philosophy

David E. Hahm in *The Origins of Stoic Cosmology* writes that after the dissolution of the universe "god is left behind in the water as the *spermatikos logos* of the cosmos.[32]" He quotes Diogenes who used the following analogy to describe the potential or seed-state of the universe: "As the seed is embraced in the seminal fluid, so god is left behind in the wet." Hahm comments:

The wet, or water, which constitutes the cosmos at this stage, is compared to the seminal fluid; and god is compared to the seed in the seminal fluid. In fact, he is called a *spermatikos logos*, a "*logos* pertaining to a seed."[33]

---

[29] Sir Ernest A. Wallis Budge, *Gods of the Egyptians*, Vol. I, pp. 310 f., quoted in Carl G. Jung, *Aion*, R. F. C. Hull, trans., 2nd ed. (Princeton, NJ: Princeton University Press, 1970), p. 207.

[30] Henri Frankfort, *Kingship and the Gods*, Samuel Noah Kramer, pref. (Chicago & London: The University of Chicago Press, 1948, 1978), p. 153.

[31] Quoted in Sir James George Frazer, *The Golden Bough*, Vol. I (New York: Macmillan Co., 1955), p. 303.

[32] David E. Hahm, *The Origins of Stoic Cosmology* (Ohio State University Press, 1977), P. 75.

[33] Ibid., p. 60.

A. K. Hunt provides the following information: "Just as the seed is contained in the seminal liquid so God, who is the *Spermatikos Logos* of the universe, remains behind in the moisture as just such an agent."

> Not only does the *Spermatikos Logos* determine the structure of the cosmos as a whole but it also contains within itself the *spermatikoi logoi* of the species of things and creatures which all develop from these as from seeds, whence in each cycle their nature is constant. ...
> We have here an hypothesis not unlike that of the modern geneticists who hold that in one tiny cell are contained all the instructions necessary for the development of the animal destined to be formed from it.[34]

In the cosmologic system of Gnostic Basilides the agent that contains the essence of all things that exist in the universe is called "the cosmic seed."

> The cosmic seed contains everything within it, [just] as the grain of mustard-seed contains everything at once within the smallest space: the roots, the stem, the branches, innumerable leaves, seeds of seeds which are produced by the plant itself, and hence a multitude of ever other seeds and other plants. Thus did ... God create a ... cosmos, by letting a single seed fall in its place, which contained the whole germ of the cosmos.[35]

With regard to Greek cosmology David E. Hahm calls attention to the fact that "The Pythagoreans seem to have thought the cosmos grew from the unit as a seed. This unit-seed began to inhale the infinite void surrounding it, and by imposing limit on it produced the cosmos." He also points out that "Pherecydes of Syros wrote a cosmogony in which the semen of Chronos played a part," and that according to the Stoic school of philosophy, founded by Zeno of Citium about 300 B.C., "In biological terms Zeus's seminal emission supplies both the creative power and the matter out of which the cosmos is made."[36]

It also has been noted that in Greek cosmology the formative and regulating principle of the universe is the *logos spermatikos*. "In order to form the world," writes Eduard Zeller, "God first transformed a part of the fiery vapour of which he consisted into air and then into water, in which he was immanent as the formative force (*logos spermatikos*)."[37]

Finally it may be added that in *Aion* Jung remarks that in the view of the Gnostic Naassenes the *logos spermatikos* or formative power of the universe is contained in the "procreative seed" of the world, and that they derive man's procreative power from this procreative seed of the universe.

---

[34] Hunt, op. cit. note 21, pp. 37-38.
[35] The cosmic system of Basilides as described by Leisegang in *Gnosis*, pp. 215 f., quoted in von Franz, op. cit. note 20, p. 399.
[36] Hahm, op. cit. note 32, pp. 65 and 62.
[37] Zeller, op. cit. note 22, p. 216.

For the Naassenes, says Hippolytus, place the "procreative nature of the Whole in the procreative seed."... But one should not overlook the fact that in reality man's procreative power is only a special instance of the "procreative nature of the Whole." "This, for them, is the hidden and mystical Logos."... Although the substance of this seed is the cause of all things, it does not partake of their nature. They say therefore: "I become what I will, and I am what I am." For he who moves everything is himself unmoved. "He, they say, is alone good."[38]

Creation from Seed in Christianity

The Christian conception of the origin of the universe is essentially identical with the *Spermatikos Logos* idea of the Stoics. It equates the cause of all things with the Word, which term is the translation of the Greek *Logos*. In many Christian texts the Word or *Spermatikos Logos* is referred to as "the good seed," and in 1 Peter 1:23 the chief apostle of Jesus calls it the imperishable seed.

In the *Epideixis* of Irenaeus by this imperishable seed or "Word of God the whole universe is ordered and disposed," and in St. John 1:1-4 it is said: "In the beginning was the Word, ... and the Word was God.... All things were made by him; and without him was not any thing made that was made. In him was life; and the life was the light of men."

The seminal Word not only gave birth to the universe, but in St. John 1:14 "the Word was made flesh," and from that point it was called "Jesus Christ," "the Son," "the Only-begotten."

The statements that "the Word was God" and "the Word was made flesh" are significant, because if the Word gave life to a human being, then it had to be the human genome. Indeed, in St. Luke 8:10-11 Jesus himself equates the Word with the seed. He makes this revelation: "Unto you it is given to know the mysteries of the kingdom of God: ... The seed is the word of God."

Allegro agrees. He writes that the term Word is simply one of the many names for God's semen. In his words: "The seed of God was the Word of God." Again he notes: "The god was the seed, the creative semen of the god was also the Word."[39]

Given that the Word means creative semen, the conclusion is inescapable that in the beginning was the creative semen, that the creative semen was God, and that all things were made by the creative semen.

Finally mention should be made of the equation of Atum-Re's name with the Word. According to Manly P. Hall Re's secret and sacred name "is equivalent to the Lost Word of Masonry." "By means of this Word," he writes, the "... priests of Isis became adepts in the use of the unseen forces of nature."

---

[38] Carl G. Jung, Aion, R. F. C. Hull, trans., 2nd ed. (Princeton, NJ: Princeton University Press, 1970), p. 201.
[39] Allegro, op. cit. note 26, pp. xii and 28.

Hall explains that the Master Mason's Word was lost when Chiram Abiff, whose name signifies "My Father, the Universal Spirit, one in essence, three in aspect," was killed for refusing to reveal "the Master's *Word*," "the password of the Master's degree."

Hall also says that in the Masonic Mysteries Chiram "is buried in the elements of creation;" that he "sets the primordial atoms in motion, and establishes order out of Chaos;" that as the third, or incarnating aspect of the Creator he is "the Master Builder who through all ages erects living temples of flesh and blood as shrines of the Most High;" that he lies waiting for rediscovery and development "as the potentiality of cosmic power within each human soul."

Hall concludes:

> As Chiram when raised from his grave whispers the master Mason's Word which was lost through his untimely death, so according to the tenets of philosophy the reestablishment or resurrection of the ancient Mysteries will result in the rediscovery of that secret teaching without which civilization must continue in a state of spiritual confusion and uncertainty.[40]

The correlation of Re's name with the Lost Word of the Freemasons is significant. It indicates that the intended objective of this secret society is to rediscover that the Great Architect of the Universe is actually the enshrined cosmic seed within each human being.

---

[40] Manly P. Hall, *The Secret Teachings of All Ages*, 19th ed. (Los Angeles: The Philosophical Research Society, Inc., 1973), pp. XLVIII, LXXVIII, and LXXX.

I pray you, look with the eyes of the mind at this little tree of the grain of wheat, regarding all its circumstances, that you may be able to plant the tree of the philosophers.

Jean d'Espagnet
Hermetic Arcanum

Chapter Three

THE COSMIC TREE

The creation stories indicate that in ancient times it was known that a direct analogy exists between the life functions of biological systems and that of the cosmic system. It was known, in other words, that just as the tree is the seed unfolded or materialized, so the universe is the cosmic seed unfolded or materialized. For this reason the image of a sacred tree was often used to illustrate the progressive development of the cosmic system from its ultimate cause. Roger Cook put it this way in *The Tree of Life*:

> The Kabbalists understood creation to be the outward manifestation of the inner world of God, and they used the image of the inverted tree to exemplify this idea. For, just as the seed contains the tree, and the tree the seed, so the hidden world of God contains all Creation, and Creation is, in turn, a revelation of the hidden world of God.[41]

In addition Cook points out that, "in the Indian scriptures, it was the banyan [tree], with its long aerial roots, which provided the underlying image of the inverted tree." He adds: "Its natural action reflected the vigorous manifestations of the sacred throughout the cosmos, from a single and transcendent source, the 'seed' of Brahman."[42]

Mircea Eliade makes the same point. He writes: "The tree came to express the cosmos fully in itself, by embodying, in apparently static form, its 'force', its life and its quality of periodic regeneration." In a word, says Eliade, "The cosmos is symbolized by a tree ..."[43]

Manly P. Hall also notes: "Several ancient peoples—notably the Hindus and Scandinavians—regarded the Macrocosm, or Grand Universe, as a divine tree growing from a single seed sown in space." Moreover, he writes, "Kapila declares the universe to be the eternal tree, Brahma, which springs from an imperceptible and intangible seed—the material monad." Finally Hall comments:

> The growth of the universe from its primitive seed may be likened to the growth of the mighty oak from the tiny acorn. While the tree is apparently much greater than its own source, nevertheless that source contains potentially every branch, twig, and leaf which will later be objectively unfolded by the process of growth.[44]

---

[41] Roger Cook, *The Tree of Life: Image for the Cosmos* (New York: Avon Books, 1974), p. 18.
[42] Ibid.
[43] Mircea Eliade, *Patterns in Comparative Religion* (New York: Meridian Book, 1965), pp. 271 and 309.
[44] Hall, op. cit. note 40, p. XCIV.

In the *Ahapramamasamyak-nama-dakini-upadesa* this genetic concept of the world is summarized as follows:

> As the seed, so the tree—
> As the tree, so the fruit.
> Looking at the whole world in this way—
> This then is relativity.

Chapter Four

THE COSMIC CYCLE

In the cosmologies that derive the universe from a perpetual seed the creation of the universe is not considered to be a true beginning, but only the beginning of one of the cosmic cycles, initiated by the germination of the immortal seed. "The Sung philosophers believed in a destruction of the world, but followed by a reconstruction," writes Alfred Forke. He adds: "Creation and destruction are repeated in regular periods."

Forke relates that when the sage Huan-lung-tse was asked to explain the origin of the world, "he once replied that it had none and another time that it had one, adding by way of explanation, 'Speaking of but one world-period it has a commencement, having in view all the world-periods, however, it has none'."[45]

In the *Srimad Bhagavatam* the following analogy is drawn from biology to illustrate the eternal cosmic cycle:

> As the spider weaves its thread out of its own mouth, plays with it, and then withdraws it again into itself, so the eternal, unchangeable Lord, who is formless and attributeless, who is absolute knowledge, and absolute bliss, evolves the whole universe out of himself, plays with it, and again withdraws it into himself.[46]

A commentary in the *Bhagavad-Gita* sums up the seed cosmology in this passage:

> Since it is subject to the eternal power of Brahman, the universe is part of a beginningless and endless process, which alternates between the two phases of potentiality and expression. When, at the end of a time-cycle, or kalpa, the universe is dissolved, it passes into a phase of potentiality, a seed-state, and thus awaits its next creation.[47]

References to the cosmic cycle are also detectable in the Egyptian myths. In his *Egyptian Ideas of the Future Life* Sir Wallis Budge writes that in the *Papyrus of Nesi Amsu* the god Neb-er-tcher makes this self-revelation: "My name is Ausares (Osiris), the germ of primeval matter.... Nothing existed on this earth then, and I made all things."[48] In referring to Osiris Pierre Grimal notes: "Osiris, then, is the seed, which dies when buried in the ground, only to be born again."[49]

---

[45] Forke, op. cit. note 6, pp. 111-113.

[46] *Srimad Bhagavatam* XI. iii, quoted in Perry, op. cit. note 1, p. 33.

[47] Swami Prabhavananda and Christopher Isherwood, trans., *The Song of God: Bhagavad-Gita*, 3rd ed. (New York: Mentor Book, 1955), p. 132.

[48] Sir Ernest A. Wallis Budge, *Egyptian Ideas of the Future Life* (London: Kegan Paul, Trench, Trubner & Co., Ltd., 1900), p. 23.

[49] Grimal, op. cit. note 15, p. 36.

With respect to this cycle of death and resurrection associated with the Egyptian gods Henri Frankfort makes the following observation: "In a curious dialogue with Osiris, Atum predicts the destruction of the world he has created and his own reversal to the shape of 'a snake whom nobody knows' ..."[50]

The reversal of the universe to a snake-state brings to mind the legendary phoenix, a bird which periodically burns itself to ashes, but through an act of asexual procreation rises youthfully to life again from the ashes of the old bird.

In *The Myth of the Phoenix* H. Van Den Broek notes: "Tacitus has a very original version of this: the phoenix impregnates the nest, and the semen gives rise to a young bird." Van Den Broek also writes:

> It seems justified to assume that the semen is to be identified with the fluid that flows from its [the old phoenix's] decomposing body and in which, according to the texts cited above, the renewal of the phoenix begins. That the semen is susceptible to this interpretation is shown by Ambrose's report: after telling of the origin from the body fluids, the Church Father says, a little further on, that the Creator willed that this bird reproduces itself from its own seed.[51]

The fact that the Egyptian myths speak of the destruction of the world and of its recreation through Atum-Re's self-fertilizing act, and also of the periodic resurrection of the phoenix from its ashes through an act of self-generation, suggests that the phoenix is an image of the cosmos. Thus it is fairly clear that the idea was not foreign to the Egyptians that the universe dies at the end of each cosmic cycle; that the seed of the universe always remains; and that when it germinates, it acts on the ashes of the previous universe, and gives life to a new universe by reducing chaos to order.

As has already been indicated, in Greek philosophy it is similarly assumed that a new creation always follows the destruction of the universe. In many cases the Greek philosophers used biological analogies for the illustration of this endless cyclic process. Evidently they were convinced that the universe is a living system, and for this reason its operation can be best explained in terms of biological systems. As David E. Hahm writes regarding Cleanthes:

> According to Stobaeus, Cleanthes reasoned that "just as all the parts of a thing grow from seeds at the fitting times, so the parts of the universe grow at the fitting times; and as *logoi* of the parts collect together into a seed ... and are again separated when the parts come into existence, so from one all things come to be, and from all one is combined."[52]

---

[50] Frankfort, op. cit. note 30, pp. 145-146.
[51] R. Van Den Broek, *The Myth of the Phoenix* (Leiden: E. J. Brill, 1972), pp. 216-217 and 188-189.
[52] Hahm, op. cit. note 32, p. 79.

Hahm provides this summary regarding the Stoic account of the periodical renewal of the cosmos:

> In the conflagration the elements of the cosmos change into fire as into a seed (sperma), so that the fire may be called the seed of the future cosmos. Thus the Stoics obviously saw in the cycle of conflagration and restoration a cycle of growth and reproduction, the cycle that every species of animal experiences. Moreover, since the cosmos grows and reproduces itself, it does not actually perish, but survives eternally. For this reason from the biological point of view, just as from the physical point of view, the cosmos must be deemed eternal.[53]

It is worth noting here that what is eternal is not the cosmos, as Hahm believes, but the seed of the cosmos. Anyway, Eduard Zeller summarizes the endless cyclic process as follows:

> After the present world-period has run its course, a conflagration will transform all things into a huge mass of fiery vapour: Zeus takes back the world into himself, to emit it again at a preordained time. Thus the history of the world is an endless succession of creation and destruction.[54]

It should be recalled here that evidence has already been given which shows that Zeus is the "seed of life." Thus it may be said that periodically the seed of life withdraws the world into itself to emit it again at the next round of creation.

Concerning the biblical story of creation the probability is that Genesis does not refer to a first creation of the universe, or to creation from nothing. Bruce Vawter writes in his book, *On Genesis: A New Reading*: "It has often been observed that the Hebrew word translated 'created' (*bara*) does not necessarily mean making something out of nothing, that it can be used (cf. Isaiah 41:20, for example) to refer to a re-creation, the remaking of something that already exists."[55]

Given the probability that the Genesis story is about the re-creation of the universe, and not about its first creation, the proponents of creation science should reexamine their absurd claim that the universe, energy, and life were suddenly created from nothing.

In the *Chandogya Upanishad* there is an argument which points out that no reasonable mind can assume creation from nothing, or *creatio ex nihilo*. It is in the form of a lecture that was given by a Hindu sage to his son, called Svetaketu. Throughout the text the maker of the universe is referred to by such names as "Existence," "the One," "the Self." The sage explained:

---

[53] Ibid., p. 194.
[54] Zeller, op. cit. note 22, pp. 216-217.
[55] Bruce Vawter, *On Genesis: A New Reading* (Garden City, NY: Doubleday, 1977), p. 38.

"In the beginning there was Existence, One only, without a second. Some say that in the beginning there was non-existence only, and that out of that the universe was born. But how could such a thing be? How could existence be born of non-existence? No, my son, in the beginning there was Existence alone—One only, without a second. He, the One, thought to himself: Let me be many, let me grow forth. Thus out of himself he projected the universe; and having projected out of himself the universe, he entered into every being. All that is has its self in him alone. Of all things he is the subtle essence. He is the truth. He is the Self. And that, Svetaketu, that art thou."[56]

Desirous of knowing more about the Self—i.e., about the hidden "subtle essence"—, Svetaketu asked for more information. The father called his son's attention to the existence of the hereditary material in seeds.

"... Bring a fruit of that Nyagrodha tree." "Here it is, sir."
"Break it."
"It is broken, sir."
"What do you see?"
"Some seeds, extremely small, sir."
"Break one of them."
"It is broken, sir."
"What do you see?"
"Nothing, sir."
"The subtle essence you do not see, and in that is the whole of the Nyagrodha tree. Believe, my son, that that which is the subtle essence—in that have all things their existence. That is the truth. That is the Self. And that, Svetaketu, that art thou."[57]

As we can see, by using the Nyagrodha tree as an analogical tool the sage is teaching his son that the universe unfolds from a seed, similarly as the Nyagrodha tree unfolds from a seed; that the universe is pervaded with the "subtle essence" of its parent seed, similarly as the Nyagrodha tree is pervaded with the "subtle essence" of its parent seed; that the universe undergoes a cycle of creation and destruction, as the Nyagrodha tree, but the "subtle essence", or seed of the universe, survives, similarly as the seed of the Nyagrodha tree survives; that the parent seed of the universe is reproduced in man, similarly as the parent seed of the Nyagrodha tree is reproduced in its fruit.

Jili, the great Sufic metaphysician, put it this way: "You must know that the Perfect Man is a copy of God." John Smith the Platonist expressed the same concept in these words: "God hath stamped a copy of his own archetypal loveliness upon the soul, that man by reflecting into himself might behold there the glory of God."

---

[56] Prabhavananda and Manchester, op. cit. note 13, pp. 68 - 69.
[57] Ibid., p. 70.

Chapter Five

THE HIDDEN GOD IN MAN

The tree symbol of the cosmos suggests that if the universe is the Tree of Life, then man is the "fruit", end product, or output of the cosmic system. It is conceivable, therefore, that man has the inherent potential to bring a universe into existence, similarly as the seeds of a tree have the inherent potential to grow into a tree. Manly P. Hall, commenting on an alchemical text, remarked:

> A philosopher might declare that a universe could be made out of a man, but the foolish would regard this as an impossibility, not realizing that a man is a seed from which a universe may be brought forth.[58]

Philip Rawson also notes in his book, *Tantra*, that the vital key to the Tantric Genesis is man, because, "figuratively speaking, the whole universe is contained within the human body. But this is something that can only be realized in a special flash of intuition." He adds: "Indian tradition has described many occasions on which some divinity has shown him or herself to followers as containing all the stars, universes, worlds, and creatures down to the minutest, within his body."[59]

In the *Bhagavad-Gita*, for example, it is related that on one occasion Krishna revealed his "Universal Form" to Prince Arjuna, whose hair stood on end when he "beheld the entire universe, in all its multitudinous diversity, lodged as one being within the body of the God of gods." Krishna explained: "0 Arjuna, I am the divine seed of all lives. In this world, nothing animate or inanimate exists without me."[60] He continued:

> Once more I shall teach you
> That uttermost wisdom: …
> Prakriti, this vast womb,
> I quicken into birth
> With the seed of all life:
> Thence, 0 son of Bharata,
> The many creatures spring.
> Many are the forms of the living,
> Many the wombs that bear them;
> Prakriti, the womb of all wombs
> And I the seed-giving Father.[61]

---

[58] Hall, 22, cit. note 40, P. CLIV.

[59] Philip Rawson, *Tantra: The Indian Cult of Ecstasy* (New York: Avon Books, 1973), P. 20.

[60] Prabhavananda and Isherwood, op. cit. note 47, p. 90.

[61] Ibid., p. 106.

"It is a great truth, which you should seriously consider," writes Paracelsus, "that there is nothing in heaven or upon the earth which does not also exist in Man, and God who is in heaven exists also in Man, and the two are but One."[62]

This "great truth," that God and man "are but One," is stated by no less an authority than Jesus Christ. In St. John 10:34 he reminds the unbelieving Jews: "Is it not written in your law, I said, Ye are gods?" The same teaching is evident in many other documents. For example Epictetus, the Greek-born Stoic philosopher, said: "You bear God about with you, poor wretch, and know it not. Do you think I speak of some external god of silver or gold?" Hermes also said: "If you possess true knowledge (gnosis), 0 Soul, you will understand that you are akin to your Creator."

In *The Early Church and the World* Cecil John Cadoux notes: "The Logos that was incarnate in Christ existed as a 'seed of reason' in every man."[63]

Allegro indicates that Jesus or the "seed of reason" represents man's seed or genome. He writes that the Hebrew original of the name Jesus, "*yehoshua*, Joshua, comes from Sumerian IA-U-ShU-A (ShUSh), 'semen, which saves, restores, heals'."[64]

Herbert Lockyer also equates Christ with the seed. He writes:

As "the Word of God" is the seed, and Christ came as "the Word of God" (John 1:1), He Himself is the Seed.... The seed we sow, then, is not only *from* Christ—it *is* Christ. Arnot expresses it, "The seed of the kingdom is Himself the King. Nor is there any inconsistency in representing Christ as the Seed while He was in the first instance also the Sower.... The incident in the synagogue at Nazareth (Luke 4:16-22) is a remarkably distinct example of Christ being at once the Sower and the Seed.... The Saviour preached the Saviour, Himself the Sower and Himself the Seed."[65]

There are many more references to the seed or hereditary nature of Jesus. For example in an early Christian text the Savior is called "Jesus Christ, the holy Seed,"[66] and John Sparrow, the English Hermetist, refers to him as "Christ Jesus, the Immortal Seed: The divine essence."[67]

On his part Jung observes that "in unconscious humanity there is a latent seed that corresponds to the prototype Jesus." He quotes Hippolytus who said that according to the Naassenes the cosmic Anthropos or cosmic Man "is the 'undivided point,' the 'grain of mustard seed' that grows into the kingdom of God. This point is 'present in the body'."[68]

---

[62] Paracelsus, quoted in Perry, op. cit. note 1, P. 927.

[63] Cecil John Cadoux, *The Early Church and the World* (Edinburgh: T. & T. Clark, 1925), p. 211.

[64] Allegro, op. cit. note 26, p. 35.

[65] Herbert Lockyer, *All the Parables of the Bible* (Grand Rapids, MI: Zondervan Publishing House, 1963), p. 177.

[66] Sir Ernest A. Wallis Budge, trans., *The Contendings of the Apostles*, Vol. II (London: Henry Frowde, 1901), p. 101.

[67] John Sparrow, quoted in Perry, op. cit. note 1, P. 314.

[68] Jung, op. cit. note 38, pp. 67 and 198.

The nature of the "cosmic Man" or "undivided point," which is present in man, is indicated by Heinrich Zimmer. He writes that in India "The Absolute is to be visualized by the concentrating devotee as a vanishing point or dot, 'the drop' (bindu), amidst the interplay of all the triangles. This Bindu is the power-point, the invisible, elusive center from which the entire diagram expands."[69]

In *Hindu World* Benjamin Walker identifies the "Bindu" or "power-point" of all creation with man's semen. He writes:

> **BINDU**, 'drop', or globule. In Hindu occultism the bindu is a metaphysical point out of time and space, the 'area' in which samadhi is experienced. It is a sacred symbol of the universe, written as a dot or the sign of zero, and in the human body symbolized and materialized in semen, in which latter sense the term is frequently used among tantriks. It is the quintessence of all manifested things.[70]

While in Hinduism the hidden God in man is the *bindu* or semen, in Egyptian tradition it is the *ka*. Jung writes: "Pharaoh was an incarnation of God and a son of God. In him dwelt the divine life-force and procreative power, the *ka*."[71] Allen Edwardes in his *Erotica Judaica* makes more than evident that the *ka* or "divine life-force" in Pharaoh stands for semen. He writes:

> Statues and bas-reliefs depict the self-begotten Supreme Being (Amen-Ra) clutching a prodigious phallus erectus and receiving the homage of Pharaohs, whom He embraces and infuses with the vital fluid (ka). The high priest of On, as avatar or embodiment of Amen, would formally and repeatedly masturbate a young Pharaoh to the point of orgasm—quickening and increasing his ka—at the same time invigorating him with divine essence by penetrating his buttocks and climaxing therein. Thus becoming a god-king imbued with hyperpotency, the new Pharaoh was periodically bound to pay sexual homage to Amen-Ra and to enact in His honor those attributes descriptive of the Omnipotent's nature.[72]

Evidently the *ka* of the Egyptians is the Spirit of God of the Hebrews. In 1 Corinthians 3:16 St. Paul asks: "Know ye not that ye are the temple of God, and that the Spirit of God dwelleth in you?" Again, in 1 Corinthians 6:19 he asks: "What, know ye not that your body is the temple of the Holy Ghost which is in you, which you have of God, and ye are not your own?"

What seems obvious is that if the *ka* or divine procreative power in Pharaoh's body is the human genome, then the Spirit of God or Holy Ghost that lives in every man is also the human genome.

---

[69] Heinrich Zimmer, *Myths and Symbols in Indian Art and Civilization*, Joseph Campbell, ed. (New York: Harper & Row, 1962), p. 147.

[70] Walker, op. cit. note 14, p. 152.

[71] Jung, op. cit. note 28, p. 259.

[72] Allen Edwardes, *Erotica Judaica* (New York: The Julian Press, 1967), p. 11.

George Perrigo Conger in his *Theories of Macrocosm and Microcosm* remarks that "the essentials of the whole [world] process are viewed as concentrated in the end-product, man."[73] "Therefore Sol is rightly named the first after God, and the father and begetter of all," reasons Gerhard Dorn in his *Physica Trismegisti*, "because in him the seminal and formal virtue of all things whatsoever lies hid."[74]

Paracelsus expresses the same view. He writes:

> Each man has the essence of God, and all the wisdom and power of the world (germinally) in himself, but ... he who does not find that which is in him cannot truly say that he does not possess it, but only that he was not capable of successfully seeking for it.

In connection with the term "essence of God" it is important to recall that Allen Edwardes has correlated the "divine essence" or *ka* with man's semen; that Jung used the word "essence" in connection with man's "seminal matter"; and that in Felix Mann's *Acupuncture* Ling Shu identifies the Spirit with the united "Life Essence (the male and female semen)."

Moreover it must be mentioned that according to the *Brihadaranyaka Upanishad* "semen is the essence of man"; that Mircea Eliade in Yoga says that "Rasa (mercury) is the quintessence of Siva," and that "It is likewise the seed of Hara (=Siva)";[75] that according to Jung Jehovah—whose name John M. Allegro shows to mean "seed of life" or semen—in the *Aquarium Sapientum* is called "universal essence"; and that Jonn Mumford writes that converted semen is "purified quintessence."[76]

Paracelsus implies that man is the seed of the universe when he says that he is the essence of all creation. He writes: "Man is a microcosm, or a little world, because he is an extract from all the stars and planets of the whole firmament, from the earth and the elements; and so he is their quintessence." Elsewhere he clearly states that man is the seed of the world, and uses the following analogies to illustrate his point:

> The whole world surrounds man as a circle surrounds one point. From this it follows that all things are related to this one point, no differently from an apple seed which is surrounded and preserved by the fruit, and which draws its sustenance from it.... Similarly, man is a seed and the world is his apple; and just as the seed fares in the apple, so does man fare in the world, which surrounds him.[77]

---

[73] George Perrigo Conger, *Theories of Macrocosm and Microcosm* (New York: Russell & Russell, 1967), p. 48.

[74] Gerhard Dorn, quoted in Jung, op. cit. note 28, p. 94.

[75] Mircea Eliade, *Yoga* (Princeton, NJ: Princeton University Press, 1969), p. 282.

[76] Mumford, op. cit. note 18, p. 30.

[77] Jolande Jacobi, ed., *Paracelsus: Selected Writings*, Norbert Guterman, trans. (Princeton, NJ: Princeton University Press, 1969), p. 38.

Many philosophers called man's attention to his divine nature. For example Plotinus claimed that "each of us is an Intellectual Cosmos," and "Democritus said man is actually a little cosmos," writes David E. Hahm.[78] Manly P. Hall also notes: "Pythagoras taught that both man and the universe were made in the image of God; that both being made in the same image, the understanding of one predicated the knowledge of the other."[79] "Man has been truly termed a 'microcosm', or little world in himself," remarks Al-Ghazali, the famous Muslim theologian, "and the structure of his body should be studied not only by those who wish to become doctors, but by those who wish to attain to a more intimate knowledge of God." Avicenna simply said: "Thou believest thyself to be nothing, and yet it is in thee that the world resides."

Christian philosophers stress the same point. For example Johannes Scotus Erigena writes: "Man is the microcosm in the strictest sense of the world. He is the summary of all existence. There is no creature that is not recapitulated in man." As St. Gregory Palamas notes: "Man, this major world in miniature, is a unified abridgement of all that exists, and the crowning of divine works." Nicholas of Cusa agrees that human nature, "embracing within itself all things, has very reasonably been dubbed by the ancients the microcosm or world in miniature." Jan Baptista van Belmont stated the same concept in these words: "Man is the mirror of the universe, and his *triple* nature stands in relationship to all things."

So far as Helena P. Blavatsky is concerned, she makes the following comment on the Jewish Kabbalistic book *Zohar*:

> Man was the store-house, so to speak, of all the seeds of life.... As En-Soph is "One, notwithstanding the innumerable forms which are in him" (Zohar, 1. 21a), so is man, on Earth the microcosm of the macrocosm. "As soon as man appeared, everything was complete ... for everything is comprised in man. He unites in himself all forms (*Zohar*, iii. 48a)."[80]

A Sufic Saying provides this conclusion: "The universe is a great man, and man is a little universe."

In the end it must be noted that according to George Perrigo Conger the philosopher's stone and the universal medicine or elixir were known as the microcosm because they were "supposed to contain the essence of all substances...."[81] "This interior microcosm," claims Jung, "was the unwitting object of alchemical research."[82]

---

[78] Hahm, op. cit. note 32, p. 63.
[79] Hall, op. cit. note 40, p. LXVI.
[80] Helena P. Blavatsky, *Anthropogenesis*, in *The Secret Doctrine*, Vol. II (Pasadena, CA: Theosophical University Press, 1970), pp. 289-290.
[81] Conger, op. cit. note 73, p. 70.
[82] Jung, op. cit. note 38, p. 164.

The Philosopher's Stone

"Our stone is called a little world, because it contains within itself the active and the passive, the motor and the thing moved," explains Philalethes. In Berthelot's *Collection des Alchimistes Grecs* this stone is identified as a "stone which is not a stone," but "has been called the seed." Jung comments: "Like a fruit, the lapis is at the same time a seed, and although the alchemists constantly stressed that the 'seed of corn' dies in the earth, the lapis despite its seedlike nature is incorruptible. It represents, just as man does, a being that is forever dying yet eternal."[83]

The Hungarian alchemist Nicolaus Melchior Szebeni describes the philosopher's stone as "our blessed gum which dissolves of itself and is named the Sperm of the Philosophers."[84] "This sperm is the Flying Stone," writes Rhodian, instructor of Kanid, King of Persia. Another Persian, Rachaidibi, calls the stone "The sperm or First Matter." "They also call it the universal Magnesia, or the seed of the world, from which all natural objects take their origin," writes the author of *The Sophic Hydrolith*.[85]

The Glory of the World says that the philosopher's stone "is lightly esteemed by the thankless multitude: but ... is very precious to the Sages." The same opinion is expressed in Deuteronomy 32:18. Its author says: "Of the Rock that begat thee thou art unmindful, and hast forgotten God that formed thee."

Israel Regardie in *The Philosopher's Stone* comments:

> Unconscious of the Stone of the Philosopher's within—no better term than Unconscious could have been devised to express this condition of ignorance—how else that mankind should defile this great seminal treasure, the seed of immortality.[86]

In order to assist the students of alchemy Angelus Silesius advises: "Man, enter into thyself. For this Philosophers' Stone is not to be found in foreign lands."

As Conger pointed out, the philosopher's stone was, among other things, also the elixir of life. Jung mentions the same correlation. He writes:

> Orthelius tells us that the philosophers have never found a better medicament than that which they called the noble and blessed stone of the philosophers ...[87]

---

[83] Jung, op. cit. note 17, p. 259.
[84] Nicolaus Melchior Szebeni, quoted in Carl G. Jung, *Psychology and Alchemy*, R. F. C. Hull, trans., 2nd ed. (Princeton, NJ: Princeton University Press, 1970), p. 401.
[85] *The Sophic Hydrolith*, quoted in Perry, op. cit. note 1, p. 787.
[86] Israel Regardie, *The Philosopher's Stone*, 2nd ed. (Saint Paul, MN: Llewellyn, 1970), p. 71
[87] Jung, op. cit. note 84, p. 428.

The Universal Medicine

Like Orthelius, Arnold of Villanova identifies the "medicament" with the "*lapis philosophorum*," i.e., with the stone of the philosophers. He writes: "The Philosopher's Stone cures all maladies.... It restores youth to the old." Paracelsus makes the same association. In his opinion it "purifies the whole body and cleanses it of all its filth by developing fresh young energies." In the following passage Paracelsus correlates the "word of God"—that means the genetic information in the human genome—with the heavenly medicine.

> All active virtues come from the word of God, and His words have such power that all nature with its forces cannot accomplish as much as a single one of His words. This divine power is the heavenly medicine; it accomplishes what no natural force can accomplish.

Gerhard Dorn calls the universal medicine *panacea*. In his words: "Within the human body is concealed a certain metaphysical substance, known to very few, which needs no medicament, being itself an incorrupt medicament."[88] He adds: "And we do not deny that our spagyric medicine is corporeal."[89]

The author of *The Glory of the World* claims that the hidden "metaphysical substance" in man is capable of transforming not only the human body, but imperfect metals as well.

> The suffering, disease, and imperfection brought not only upon men, but also upon plants and animals, by the fall of Adam, found a remedy in that precious gift of Almighty God, which is called the Elixir, and Tincture, and has power to purge away the imperfections not only of human, but even of metallic bodies; which excelles all other medicines, as the brightness of the sun shames the moon and the stars.[90]

Swami Sivananda calls this gift of God "divine elixir that makes one Godlike." Regarding the nature of the "divine elixir" Jonn Mumford writes: "Semen, or Bindu, is held to be the true elixir of life by Yoga and Tantric schools alike."[91] Allegro explains: "Since all life derives from the divine seed, it follows that the most powerful healing drug would be the pure, unadulterated semen of the god." In addition Allegro mentions that "human semen was a cure for scorpion stings, according to Pliny ..."[92]

---

[88] Gerhard Dorn, quoted in Ibid., p. 269.
[89] Gerhard Dorn, quoted in Jung, op. cit. note 28, p. 465.
[90] *The Glory of the World*, quoted in Perry, op. cit. note 1, p. 610.
[91] Mumford, op. cit. note 18, p. 22.
[92] Allegro, op. cit. note 26, pp. 34 and 56.

It is also on record that it came to the attention of St. Epiphanius, the bishop of Constantia, that a "Gnostic sect sacramentally administered semen to its members."[93] The same practice is recorded about the Manicheans and the Albigenses. As Ernest Crawley writes in *The Mystic Rose*:

> The Manicheans sprinkled their eucharistic bread with human semen, a custom followed by the Albigenses. Human semen, as medicine, is used by many peoples, as by the Australians, who believe it an infallible remedy for severe illness. It is so used in European folk-custom, where we also find it used as a love-charm, on the principle of transmission of qualities.[94]

Evidently by Australians Crawley means the aborigines. In connection with these people Géza Roheim observed that they administer not only blood to the sick to strengthen them, but "Seminal fluid is used for the same purpose."[95]

According to all signs this practice was universal because it can be detected in the folk-medicine and traditions of most societies. Just to mention a few more examples, Reichel-Dolmatoff observed many times among the Desana of the Northwest Amazon that things they associated with semen—such as crystalline substances, honey or starch—were used for the curing of diseases.[96]

In *The Tao of Sex* Ishihara and Levy write that according to the text called *The Secrets of the Jade Bedroom* the Woman Plain gave the following advice to the Yellow Emperor when he complained about losing his potency:

> By gathering in your overflowing semen and by taking the fluid into your mouth, the semen's life-force returns and fills up your brain.... Your body ... can then be preserved. If your orthodox life-force is moribund within, every illness will disappear.[97]

In *The Secret of the Golden Flower* Richard Wilhelm comments on the *Book of the Elixir of Life*, an old Chinese alchemical text, which is relevant here.

> An ancient adept said: 'Formerly, every school knew this jewel, only the fools did not know it wholly.' If we reflect on this we see that the ancients really attained long life by the help of the seed-energy present in their own bodies, and did not lengthen their years by swallowing this or that sort of elixir.... The holy and wise men have no other way of cultivating their lives except by destroying lusts and safeguarding the seed. The accumulated seed is transformed into energy, and the energy, when there is enough of it, makes the creatively strong body.[98]

---

[93] Francis King, *Sexuality, Magic and Perversion* (Secaucus, NJ: Citadel Press, 1974), p. 173.

[94] Ernest Crawley, *The Mystic Rose* (London: Watts & Co., 1932), p. 100.

[95] Roheim, op. cit. note 24, p. 112.

[96] Reichel-Dolmatoff, op. cit. note 2, pp. 98-103 and 152.

[97] Akira Ishihara and Howard S. Levy, *The Tao of Sex* (Han Row Books, 1970), pp. 26-27.

[98] The *Tai I Chin Hua Tsung Chih*, in Richard Wilhelm, trans. and expl., *The Secret of the Golden Flower: A Chinese Book of Life*, Carl G. Jung, forew. and comm. (New York: Harvest Book, 1962), pp. 62-63.

The alchemist William Maxwell provides this neat summary of the subject: "The universal remedy is no other than the vital spirit strengthened in a suitable subject."[99]

---

[99] William Maxwell, quoted in Regardie, op. cit. note 86, p. 91.

Cosmology has much in common with religion; both rely on a very small measure of information and a very large measure of belief.

Astronomer Geoffrey Burbidge
Time, December 27, 1976

Chapter Six

CREATION FROM SEED VS. BIG BANG

At first the theory may seem incredible that the universe originates from a seed. But the mathematical Big Bang model—that traces the origin of the universe and life back to the explosion of a super-dense, dimensionless singularity—sounds even more incredible. In order to provide a basis for comparison, here is a description of the Big Bang model by the Soviet astronomer Iosif Shklovsky:

> In the first instants of its history the Universe was an inconceivably dense "droplet" the size of an atom, which, in some unthinkable way, contained the entire matter of the future Universe. Of course, it is extremely difficult to conceive that this tremendous mass (if estimated in tons, its weight is something like 10 plus fifty zeroes) was compressed into such a microscopic particle, but it is nevertheless a fact.
>
> Then, 20 thousand million years ago, for reasons which we are ignorant of, this "droplet" exploded. The explosion was so monstrously enormous that the matter contained in it is still being scattered in different directions at a tremendous speed.[100]

Scientists speculate that as the temperature decreased, the contraction of gas clouds—through some still undetermined mechanism—led to the formation of superclusters, galaxies, solar systems, and planets. In their view the Big Bang brought into existence not only the heavenly bodies, but also a series of random chemical reactions that resulted the accidental formation of a "primordial soup" of organic molecules in the oceans of the prevital earth. From this soup, scientists believe, came into being all the highly complex life forms, including man. Thus according to this evolutionary concept life is the product of a pure accident in evolution, and it might just as well not have appeared.

Shklovsky admits that "it takes an extremely rare coincidence of a tremendous number of exceptionally favourable factors to trigger off the process leading to the origin of life." He adds:

> More than that, we still cannot say clearly and precisely what circumstances led to the origin of life on our planet. There is a vast abyss between the chemical compounds necessary for the origin of life and the living organism, however simple, consisting of these compounds. Even the most primitive bacterium is a miracle, if only because it is a part of the evolutionary process which was crowned by homo sapiens.[101]

---

[100] Iosif Shklovsky, quoted in G. Polski, "Are we alone in the Universe?", Sovietskaya Rossiya, trans. and repr. in Moscow News, No. 47, 1978, p. 11.
[101] Ibid.

Harold F. Blum, one of evolution's most prominent advocates, put the problem of life's assumed origin this way in his book, *Time's Arrow and Evolution*:

> The riddle seems to be: *How, when no life existed, did substances come into being which today are absolutely essential to living systems yet which can only be formed by those systems?*[102]

Another riddle evolutionists must face is the existence of only *one* genetic code in terrestrial life. If it is true that the living creatures came into being in some great "primeval soup," then it is strange that organisms with a number of different genetic codes do not exist.

Nobel laureate Francis Crick, co-discoverer of the DNA structure, and his biochemist colleague, Leslie Orgel, see the existence of a single code entirely compatible with their idea that life did not arise spontaneously on earth from some prebiotic soup, but far out in space, and that this planet had been deliberately seeded with life by a distant civilization.

As Francis Crick points out in his article, "Seeding the Universe," this idea is a modern version of the old panspermia ("seeds everywhere") hypothesis which postulates that the germs of life were accidentally brought to earth over interstellar distances by comets or cosmic particles under radiation pressure. "But Orgel and I suggested," writes Crick, "that the microorganisms more probably had arrived in a space vehicle 'homed in' on Earth by a distant civilization. We called this concept directed panspermia."[103]

Critics of the panspermia hypothesis say that even if this idea is correct, it avoids rather than solves the question of the origin of life on this planet by transferring the problem to some other heavenly body. Moreover they doubt that any microorganism could survive the long journey across space, unless it arrived to earth in a space capsule.

These problems associated with the evolutionary Big Bang theory indicate that those who attempt to derive the universe and life purely from a preexistent super-dense mass of matter are about as much in error as those who attempt to derive all matter, the universe and life purely from some animating principle or God. The first group, by denying the preexistence of an animating principle, has no credible explanation for the emergence of life. The second group, on the other hand, by denying the preexistence of the material elements, has no credible explanation for their coming into being from an animating principle.

These considerations suggest that the existence of the particles of matter and life in the universe can be explained only if it is assumed that both life and the particles of matter are without beginning. It also seems evident that it was the animating principle or perpetual seed of the universe—instead of a primeval explosion or Big Bang—that acted upon the particles of chaos to bring the structure of the universe into existence.

---

[102] Harold F. Blum, *Time's Arrow and Evolution*, 3rd ed. (Princeton, NJ: Princeton University Press, 1968), p. 164.
[103] Francis Crick, "Seeding the Universe," *Science Digest*, November 1981, p. 82.

Chapter Seven

SIGNIFICANCE OF THE SEED COSMOLOGY

Tolstoy, the Russian novelist, remarked regarding the state of science that "What is called science today consists of a haphazard heap of information, united by nothing." This is as true today as it was in Tolstoy's time. The sciences are united by nothing because the revelation that the existing highest form of life constitutes the seed of the universe is being scoffed at by most scientists. Because this Universal Common Ancestor—that holds all creation together—is being derided, the evils of the world, its advancing materialism, its ideological differences, and its narrow-minded feelings of racism, sexism, and so forth, continue to undermine relationships on all levels. In his encyclical letter Pope Pius XI warned:

> In place of moral laws, which disappear together with the loss of faith in God, brute force is imposed, trampling on every right. Old time fidelity and honesty to conduct and mutual intercourse extolled so much even by the orators and poets of paganism, now give place to speculations in one's own affairs as in those of others without reference to conscience. In fact, how can any contract be maintained, and what value can any treaty have, in which every guarantee of conscience is lacking? And how can there be talk of guarantees of conscience, when all faith in God and all fear of God has vanished? Take away this basis, and with it all moral law falls, and there is no remedy left to stop the gradual, but inevitable destruction of peoples, families, the State, civilization itself.[104]

The Church is clearly aware of the problems. Yet it fails to make God's reality credible. And if God's reality is not credible, the doctrine cannot serve effectively as a deterrent to acts which are contrary to the laws of God. As the Unitarian clergyman William Ellery Channing appropriately remarked:

> As yet Christianity has done little, compared with what it is to do, in establishing the true bond of union between man and man. The old bonds of society still continue in a great degree. They are instinct, interest, force. The true tie, which is mutual respect, calling forth mutual, growing, never-failing acts of love, is as yet little known. A new revelation, if I may so speak, remains to be made; or rather, the truths of the old revelation in regard to the greatness of human nature are to be brought out from obscurity and neglect.[105]

---

[104] Pius XI, "Encyclical: Caritate Christi Compulsi," May 1932, in John Eppstein, *The Catholic Tradition of the Law of Nations* (London: Burns Oates & Washbourne Ltd., 1935), P. 221.
[105] William Ellery Channing, quoted in George Seldes, comp., *The Great Quotations*, (Secaucus, NJ: Castle Books, 1977), p. 151.

The primary goal of this seed cosmology is to bring out from obscurity and neglect the teaching or revelation that man constitutes the seed of the universe, or the cosmic system's input and output, similarly as a seed constitutes the tree system's input and output.

This true doctrine or theory of creation has always existed in the world, but remained in obscurity throughout the ages. However in St. John 16:25 Jesus has promised: "... the hour is coming when I shall no longer speak to you in figures but tell you plainly of the Father."

The knowledge that the universe is the phenotype of man's genotype, and that we are the reproductions of that universal Common Ancestor, is significant. It provides a unified theory of the universe with the potential to bring order and meaning to a great number and variety of phenomena that otherwise would remain disconnected and unintelligible. In other words it provides mankind with a point of unity that makes the integration of all peoples possible. However that transformation in the relations of peoples, as well as in the relations of nations, will not take place unless all the nations and all the members of human society will have the intimate conviction that, descended as they are from the seed of the universe, they are all of one common genetic stock, all children of the same universal Common Ancestor or parent seed, and all parts of the same human family.

The other practical significance of the seed cosmology is that it provides the theoretical foundation for the extraction of the cosmic seed's inexhaustible life-energy. Man, in other words, will be again in the position to utilize the cosmic seed in his body for purposes which may be called "magical."

The English journalist and historian Trevor Ravenscroft in *The Spear of Destiny* shows that Hitler and a number of Nazi scientists—all of them members of the secret society called "Vril" or "Luminous Lodge"—were carrying out research in this field. Evidently they intended to rediscover and to master the energy of the cosmic seed hidden in us in order to master the world.

According to Ravenscroft the outside world learned about these activities in 1933 when the rocket scientist Willi Ley defected from Nazi Germany. "Ley's reports were quite correct," he writes, "that the Initiates of the Vril spent untold hours in silent contemplation of seeds, leaves, blossoms and fruits, even apples cut in half!"[106]

> Well versed in the whole story and historical background of *Parsival*, Adolf Hitler was aware of the medieval doctrine of correspondences between Macrocosm and Microcosm in which man is spoken of as the concentrated image of the whole Cosmos. And how the creative principle (Word) of the Universe has been implanted in man and comes to expression in the tremendous faculty of human speech.[107]

---

[106] Trevor Ravenscroft, *The Spear of Destiny* (New York: Bantam Books, 1974), pp. 244-245.
[107] Ibid., p. 181.

Although the Nazis have failed to identify the creative power of the universe with man's genetic constitution, and consequently never had a chance to exploit it to their advantage, Ravenscroft writes that they were aware of traditions which told them that in ancient times seminal power was utilized for practical purposes. He mentions that Karl Haushofer, the founder of Geo-Politics and the leading figure of the Vril society, provided Hitler with a remarkable description of the magical powers that were employed by the legendary Atlanteans.

> Many of the valid descriptions which Karl Haushofer repeated to Adolf Hitler regarding the conditions of life in ancient Atlantis must appear fantastic and startling to minds which have been conditioned uncompromisingly to the sterile and petrified concepts of modern materialism....
> Atlantean man was not the crude and primitive creature of the kind of pre-history erroneously envisaged by modern science and contemporary anthropologists.... Atlantean scientists discovered means of extracting life-power from seeds and made these forces available to large-scale commercial enterprises which arose throughout the continent. Means of transport not only included huge power-impelled ships but also flying craft with various types of sophisticated steering mechanisms.[108]

The Magic Power of Seed Power

The claim that Atlantean scientists utilized seed power for a variety of purposes is significant because it suggests that what is known today as "magic" power is actually seed power. There is another point in support of the notion that magic power is seed power. It has been noted by a number of people that the magicians practiced sexual continence in order to accumulate the "semen treasure" in their bodies. "According to the Kabi and Wakka," writes Géza Roheim about the aborigines, "a magician is a man who is full of magic stones (kundir-bongan) and these magic stones confer an extraordinary degree of vitality on the man who possesses them." Roheim adds: "The magical use of semen is prominent in New Guinea."[109] "Similarly the magicians of the West traditionally have abstained from sexual intercourse before the commencement of and during their magical operations," write Edwardes and Masters in *The Cradle of Erotica*.[110] A magus, in a word, was a man full of semen that enabled him to utilize his accumulated seminal power for such purposes as levitation, telepathy, or healing.

---

[108] Ibid., pp. 235-236.
[109] Roheim, op. cit. note 24, pp. 302 and 443.
[110] Allen Edwardes and R. E. L. Masters, *The Cradle of Erotica* (New York: The Julian Press, 1963), p. 114.

According to the Bible the Prophet Elisha demonstrated his capabilities in all three of these magical operations. In 2 Kings 6:6 he "made the iron float" when one of his disciples accidentally dropped a borrowed axe-head into the river Jordan; in 2 Kings 4:32-35 he restored the dead child of a woman to life; and in 2 Kings 6 there is a story about Elisha's telepathic abilities.

The story of this intelligence-gathering activity, carried out by Elisha for the king of Israel, runs as follows:

> Once when the king of Syria was warring against Israel, he took counsel with his servants, saying, "At such and such a place shall be my camp." But the man of God [Elisha] sent word to the king of Israel, "Beware that you do not pass this place, for the Syrians are going down there." And the king of Israel sent to the place of which the man of God told him. Thus he used to warn him, so that he saved himself there more than once or twice.
>
> And the mind of the king of Syria was greatly troubled because of this thing; and he called his servants and said to them, "Will you not show me who of us is for the king of Israel?" And one of his servants said, "None, my lord, 0 king; but Elisha, the prophet who is in Israel, tells the king of Israel the words that you speak in your bedchamber."

The Telepathy Theory

It may or may not be believable that Elisha had the supernormal capability to intercept and to monitor the most sensitive communications of the king of Syria. But if the universe indeed has its origin in a cosmic seed, then a communication network, universal and genetic in origin, must interconnect all parts of the cosmic system. Consequently it is conceivable that if an individual finds ways to tune his brain waves into that subtle and omnipresent information network, then he will be able to seek out specific targets for interception, surveillance, or even for communication.

Rufus M. Jones in his *Studies in Mystical Religion* indicates that the secret method for the tapping of another person's mind involves the purification of the human system. He quotes a text from the oracle of the Montanist prophetess Priscilla. It states that "purity unites (with the Spirit), and they (the pure) see visions, and bowing their faces downward, they hear distinct words spoken."[111]

Here the implication seems to be that those who are pure in spirit are able to link their mental powers with the universe's World Spirit, seminal essence, or quintessence. As a result of that link they receive information from the Supreme Intelligence. Thus apparently human beings have a similar relationship to the cosmic seed or Supreme Intelligence than personal computers connected to the data base of a central computer. Only those individuals who know how to access

---

[111] Quoted in Rufus M. Jones, *Studies in Mystical Religion* (London, 1909), p. 52.

the central computer may share information with the data base. In like manner only those souls who know how to access the "Soul" of the universe may share information with the Supreme Intelligence. By comparison the ways and means of the National Security Agency (NSA) still seem to be in the cradle.

With respect to telepathy Ferdinand Ossendowski quotes in his book the librarian of a Mongolian lamasery. According to that lama from his subterranean kingdom, called Agharti, "the King of the World is in contact with the thoughts of all the men who influence the lot and life of all humankind.... He realizes all their thoughts and plans. If these be pleasing before God, the King of the World will invisibly help them; if they are unpleasant in the sight of God, the King will bring them to destruction. This power is given to Agharti by the mysterious science of 'Om,' with which we begin all our prayers."[112]

In this text, no doubt, the key word is the syllable OM. It is also known as the "seed syllable" or "Word of Power." In the Bhagavad-Gita Krishna makes the following identification: "... among words, I am the sacred syllable OM...."[113] As Krishna is also the "divine seed of all lives," the sacred syllable OM evidently stands for the "divine seed" or creative seminal power of the universe. Out of this follows that the mysterious science of OM—that enables the King of the World to contact and to control his subjects—is actually an advanced form of the science of genetics.

Strategic Implications

When scientists and governments are faced with the subject of flying saucers or unidentified flying objects (UFOs), in most cases they go to great lengths to deny the possibility of their existence. On 5 April 1966, for example, the Associated Press reported that Secretary of the Air Force Harold Brown (he later became Secretary of Defense) assured the House Armed Services Committee of the U.S. Congress that "there was no evidence that the earth ever had been visited by strangers from outer space." Dr. Brown claimed that the over 10,000 unidentified flying object reports on record with the Air Force were "easily explained." He attributed the sightings of peculiar bright objects flying through the airspace of the United States to "marsh gases, pranks, planets, comets, meteors, fireballs, auroral streamers," and you name it. But even if there is no so-called "scientific evidence" in support of UFOs, to keep a closed mind to the possibility that we are not alone in the universe may prove to be counterproductive and even a threat to national security. The "unscientific" evidence is simply too convincing in favor of UFOs, so they should not be ignored—and those of us who follow the activities of the intelligence agencies know that UFOs are not ignored.

Whatever the intelligence agencies may tell the public, the British writer Desmond Leslie quotes a passage from P. Chandra Roy's translation of the

---

[112] Ferdinand Ossendowski, *Beasts, Men and Gods* (E. P. Dutton & Company, 1922), p. 309.

[113] Prabhavananda and Isherwood, op. cit. note 47, p. 89.

ancient Hindu text *Drona Parva* saying that in order to travel in the sky, "We shall build a vimana of great power.... All speeches and sciences were gathered together within it.... And the syllable OM placed before that car made it exceedingly beautiful. When it set out, its roar filled all points of the compass."[114]

In this case, just as in the case of telepathy, again the power is given by the science of OM. Thus here the seminal power of the universe is indirectly identified as the power source of the vimana.

Elsewhere in the *Drona Parva* "Cukra, riding in that excellent vimana—which was powered by Celestial Forces—, proceeded for the destruction of the Triple City.... He flings a missile which contained the Power of the Universe, at the Triple City.... Smoke, looking like ten thousand suns, blazed up in splendour."[115]

In this age, when the strategic missiles carry nuclear warheads, the Power of the Universe is associated with nuclear power. But in ancient times the supreme power was associated with the "essence of the infinite universe," which is "Brahman, the ultimate Cosmic Force," writes Lewis Browne in *The World's Great Scriptures*.[116] This can only mean that the Power of the Universe contained in the missile had to be seed energy, because Brahman—as it was pointed out earlier—is identified with the seed of the universe, i.e. with the maker of the universe.

We may infer from all this that evidently Rudolf Steiner is not fantasying when in his *Cosmic Memory* (written 1904-1908) he tells us that Atlantean scientists discovered means of extracting life-power from seeds for the propulsion of ships and flying craft.

Be as it may, the effect of the divine weapon on human beings is indicated by Flavius Philostrathus. In his book he describes the journey of Apollonius of Tyana and of his companion, Damis, to a Greek-speaking community of sages in India. These wise men—who reportedly possessed formidable weapons—also had the faculty of prescience and of levitation, and claimed to be omniscient and gods.

> Damis says that the hill on which the Sages dwell rises above the plain as high as the Acropolis at Athens, and that it is defended in the same way by a natural encircling cliff, in which are visible everywhere the impressions of cloven hoofs, and side and front faces, and here and there what look like the backs of falling men. For when Bacchus was besieging the hill with Hercules, he ordered his satyrs to assault it, thinking them strong enough to shake it, but they were smitten by the thunderbolts of the Sages, and were hurled in all directions, and after their attack had been repulsed their shapes remained imprinted on the rocks.[117]

---

[114] Desmond Leslie and George Adamski, *Flying Saucers Have Landed*. (New York: The British Book Centre, 1953), pp. 106-107.
[115] Ibid., p. 99.
[116] Lewis Browne, op. cit. note 16, p. 58.
[117] Flavius Philostratus, *Life and Times of Apollonius of Tyana*, Charles P. Eells, trans. (Stanford University, CA: Stanford University Press, 1923), p. 69.

Significantly these highly advanced wise men considered the "ether" to be the source of immortality and the animating principle of the universe. When during his stay with the sages Apollonius asked Iarchas—the person who presided over the sages—what they thought the universe consisted of, he answered: "Of the elements."

Apollonius then asked which element existed first, and Iarchas said: "They all began simultaneously, for no living thing is born piecemeal." "Do you then think the universe to be a living thing?" asked Apollonius; and Iarchas replied: "Yes, if you rightly understand it, for it gives birth to all living things." "Should we call it female, or both male and female?" asked Apollonius. "It is of both sexes," said Iarchas, "for it is self-impregnated, and acts as both father and mother in creating life, and its desire for itself exceeds all other passion of separate beings for each other, so that each part of it unites and harmonizes with the rest. Nor is there anything unreasonable in its union with itself.[118]

A review of those gods or mythological figures who rightly understood the universe shows that they were intimately linked to flying vehicles which may be called space-planes in our time. For example the Patriarch Enoch, being found righteous in observing the laws of God, was taken to heaven without tasting death. As Louis Ginzberg writes in *The Legends of the Jews*:

A few days yet Enoch spent among men, and all the time left to him he gave instruction in wisdom, knowledge, God-fearing conduct, and piety, and established law and order, for the regulation of the affairs of men. Then those gathered near him saw a gigantic steed descend from the skies, and they told Enoch of it, who said, "The steed is for me, for the time has come and the day when I leave you, never to be seen again." So it was. The steed approached Enoch, and he mounted upon its back, all the time instructing the people, exhorting them, enjoining them to serve God and walk in His ways.... On the sixth day of the journey, he said to those still accompanying him, "Go ye home, for on the morrow I shall ascend to heaven, and whoever will then be near me, he will die."...
    On the seventh day Enoch was carried into the heavens in a fiery chariot drawn by fiery chargers. The day thereafter, the kings who had turned back in good time sent messengers to inquire into the fate of the men who had refused to separate themselves from Enoch, for they had noted the number of them. They found snow and great hailstones upon the spot whence Enoch had risen, and, when they searched beneath, they discovered the bodies of all who had remained behind with Enoch. He alone was not among them; he was on high in heaven.[119]

---

[118] Ibid., p. 82.
[119] Louis Ginzberg, *The Legends of the Jews*, Henrietta Szold, trans. (Philadelphia: The Jewish Publication Society of America, 1909, 1913), pp. 129-130.

"Another legend," writes the Reverend William J. Deane, "... recounts how at Moses' death a bright cloud so dazzled the eyes of the bystanders that they saw neither when he died nor where he was buried."[120] In 2 Kings 2:11 "a chariot of fire, and horses of fire" translated Elijah to heaven. In another description of Jehovah's chariot "The spirit was its driving force. As it was shown before, the word "spirit" is a synonym for the seminal power of the universe.

In one of the Chinese legends a fiery dragon transported the Yellow Emperor to heaven. He was by no means a virgin, but he was a model of perfection, and "retained his semen while having intercourse with twelve hundred women."[121]

Krishna, similarly to Elijah, abstained from sexual intercourse. He used the celestial bird Garuda to ascend to heaven. In the *Mausala Parva* "The discus of Shri Krishna ascended to heaven, and His famous horses fled away with His celestial car."[122] Apparently the discus of Krishna was a sort of advanced weapon. When the wicked King Sisupala insulted and threatened Krishna, "the god said simply, 'Now, the cup of your misdeeds is full.' Immediately, the divine weapon, the flaming discus, rose behind Krishna and, traversing the air, fell on Sisupala's helmet and cleft him from head to foot."[123]

Regarding Gautama Buddha it is recorded that he walked upon the waters and levitated himself to the other bank of the river Ganges "on a path through the Sky."[124] In the *Digha Nikaya* he "rose in the air, projecting flames" before a large number of people. But most importantly he visited heaven several times. On one occasion, when he was reincarnated as the devoted King Nimi, the company of gods wished to see him, and they turned to Sakka, the king of the gods, to send for him.

> Sakka consented, and sent Matali: "Friend Matali, yoke my royal car, go to Mithila [city], place King Nimi in the divine chariot and bring him here." Matali obeyed and departed.... So it was the holy day of the full moon: King Nimi opening the eastern window was sitting on the upper floor, surrounded by his courtiers, contemplating virtue; and just as the moon's disk rose in the east this chariot appeared. The people had eaten their evening meal, and sat at their doors talking comfortably together. "Why, there are two moons today!" they cried. As they gossiped, the chariot became plain to their view. "No, it is no moon," they said, "but a chariot!" In due course there appeared Matali's team of a thousand thoroughbreds, and the car of Sakka, and they wondered whom that could be for? Ah, their king was righteous; for him Sakka's divine car must be sent; Sakka must wish to see their king....

---

[120] William J. Deane, *Pseudepigrapha* (Edinburgh: T. & T. Clark, 1891), p. 99.

[121] Ishihara and Levy, op. cit. note 97, pp. xi -xii.

[122] Quoted in Annie Besant, *The Story of the Great War* (London: Theosophical Publishing Society, 1899), p. 260.

[123] Grimal, op. cit. note 15, p. 243.

[124] *Lalita Vistara* XXVI, quoted in Perry, op. cit. note 1, p. 942.

As the people talked and talked, swift as the wind came Matali, who turned the chariot, and brought it to rest out of the way by the still of the window, and called on the king to enter....

The king thought, "I shall see the gods' dwelling-place, which I never have seen ...," so he addressed his women and all the people, and said—"In a short time I shall return: you must be watchful, do good and give alms." Then he got into the car.

[King Nimi was taken to the hall of the gods and stayed there for seven days.] Then the king took leave of Sakka, saying that he wished to go to the world of men. Then Sakka said, "Friend Matali, take King Nimi at once to Mithila." He got ready the chariot; the king exchanged friendly greetings with the company of gods, left them and entered the car. Matali drove the car eastwards to Mithila. There the crowd, seeing the chariot, were delighted to know that their king was returning. Matali passed round the city of Mithila right-wise, and put down the Great Being at the same window, took leave, and returned to his own place.[125]

Perhaps the most astonishing descriptions of fiery chariots and advanced weapon systems come from Ethiopic manuscripts and from the apocryphal writings of Christianity. Regarding the apocryphal texts Montague Rhodes James, the translator of these documents, writes: "An apocryphal book was—originally—one too sacred and secret to be in every one's hands: it must be reserved for the initiate, the inner circle of believers."[126]

The type of specific information the authorities considered to be "too sacred and secret" one can only guess. It seems certain, however, that they indicate the existence of a highly advanced civilization and technological power that covertly exercises influence over the affairs of humanity. To give an example, in *The Gospel of Bartholomew* the angel Gabriel announced the following to Mary, a sixteen-year-old temple virgin: "Three years more, and I will send my word [i.e., seed] and you shall conceive my son, and through him the whole world shall be saved." When Joseph and the pregnant Mary were going to Bethlehem, an angel (described in *The Gospel of Pseudo-Matthew* as "a beautiful youth, clothed in white raiment") commanded Mary to "go into a recess under a cavern, in which there never was light.... And when the blessed Mary had gone into it, it began to shine with as much brightness as if it were the sixth hour of the day.... And there she brought forth a son, and the angels surrounded Him, when He was being born." At this time Joseph was returning with two midwives, Zelomi and Salome. As they "approached the cave, they saw a luminous cloud hanging like a shining curtain over the entrance. Zelomi was awed by the sight." Joseph urged the midwives to enter the cave. When they did so, there stood by Salome "a young man in shining garments, saying: Go to the child, and adore Him...."

---

[125] E. B. Cowell, ed., *The Jataka Or Stories of Buddha's Former Births*, E. B. Cowell and W. H. D. Rouse, trans., Vol. VI (Cambridge: University Press, 1907), pp. 53-68.
[126] Montague Rhodes James, trans., *The Apocryphal New Testament* (Oxford: Clarendon Press, 1924, 1960), p. xiv.

Moreover, a great star, larger than any that had been seen since the beginning of the world, shone over the cave from the evening till the morning. And the prophets who were in Jerusalem said that this star pointed out the birth of Christ, who should restore the promise not only to Israel, but to all nations.

According to an Ethiopian Christian manuscript when the magi or wise men saw the star, "they rejoiced, for its appearance was different in many respects from that of the other stars."

Now its appearance was this: That star had the form of a virgin embracing a child in her bosom, and it travelled from left to right, and it travelled by day, and disappeared by night. When the wise men travelled, the star travelled; and when they stood still it stood still. And it was visible to them in one place, and was hidden from them in another.

And ... that star which they had seen in the East guided them until it brought them to the cave, and it stood over the cave where the Child was; and when they saw Him they rejoiced with great joy.[127]

The author of *The Life of John According to Serapion* relates that the child Jesus, when he was living with his parents in Egypt, had the supernormal faculty to perceive that his mother's kinswoman, the old Elizabeth—who was wandering in the desert with the child John—has passed away.

When the Virgin heard this she began to weep over her kinswoman, and Jesus said to her: "Do not weep, 0 my Virgin mother, you will see her in this very hour." And while he was still speaking with his mother, behold a luminous cloud came down and placed itself between them. And Jesus said: "Call Salome and let us take her with us." And they mounted the cloud which flew with them to the wilderness of Ain Karim and to the spot where lay the body of the blessed Elizabeth, and where the holy John was sitting.

The Saviour said then to the cloud: "Leave us here at this side of the spot." And it immediately went, reached that spot, and departed. Its *noise* [my emphasis], however, reached the ears of Mar John, who, seized with fear, left the body of his mother. A voice reached him immediately and said to him: "Do not be afraid, 0 John.... I am your kinsman Jesus, and I came to you with my beloved mother in order to attend to the business of the burial of the blessed Elizabeth, your happy mother....

And Jesus Christ and his mother stayed near the blessed and the holy John seven days....

Then Jesus Christ said to his mother: "Let us now go to the place where I may proceed with my work."... And they mounted the cloud, and

---

[127] Sir Ernest A. Wallis Budge, trans., *The Book of the Saints of the Ethiopian Church*, Vol. II (Cambridge: University Press, 1928), pp. 421-423.

John looked at them and wept, and Mart Mary wept also bitterly over him, saying: "Woe is me, 0 John, because you are alone in the desert without anyone...."

And Jesus Christ said to her: "Do not weep over this child, 0 my mother. I shall not forget him." And while he was uttering these words, behold the clouds lifted them up and brought them to Nazareth.[128]

Years later, relates the Apostle Peter, Jesus asked his disciples to go with him to the Mount of Olives.

And his disciples went with him, praying. And behold there were two men there, [talking with Jesus concerning his departure from this world,] and we could not look upon their faces, for a light came from them, shining more than the sun, and their raiment also was shining, and cannot be described, and nothing is sufficient to be compared unto them in this world ..., for their aspect was astonishing and wonderful.... And when we saw them on a sudden, we marvelled. And I drew near unto the Lord God Jesus Christ and said unto him: 0 my Lord, who are these? And he said unto me: They are Moses and Elias....

And behold, suddenly there came a voice from heaven, saying: This is my beloved Son in whom I am well pleased: he hath kept my commandments. And then came a great and exceeding white cloud over our heads and bare away our Lord and Moses and Elias. And I trembled and was afraid: and we looked up; and the heaven opened and we beheld men in the flesh, and they came and greeted our Lord and Moses and Elias and went into another heaven.[129]

And the disciples followed him with their eyes, and none of them spoke, until he reached the heaven, but they were all in great silence. This now came to pass on the 15th of the moon, on the day on which it becomes full in the month Tybi....

While they ... wept to one another, then the heavens opened, about the ninth hour of the following day, and they saw Jesus descend, shining very bright, and the light in which he was was beyond measure....

Then Jesus drew to himself the splendour of his light; and when this had come to pass all the disciples took courage, stood before Jesus, and all fell down together and worshipped him, rejoicing with great joy.[130]

On their way down from the Mount of Olives, according to St. Matthew 17:9, Jesus charged his disciples: "Tell the vision to no man, until the Son of man is raised from the dead."

---

[128] Edgar Hennecke and Wilhelm Schneemelcher, eds., *New Testament Apocrypha*, Vol. 1, R. McL. Wilson, trans. (Philadelphia: The Westminster Press, 1963), pp. 414-417.

[129] James, op. cit. note 126, pp. 518-519.

[130] Hennecke and Schneemelcher, op. cit. note 128, pp. 254-256.

At Christ's resurrection the soldiers guarding his tomb "saw the heavens opened and two men come down from there in a great brightness and draw nigh to the sepulchre. That stone which had been laid against the entrance to the sepulchre started of itself to roll and gave way to the side, and the sepulchre was opened, and both the young men entered in." Soon thereafter the soldiers "saw again three men come out from the sepulchre, and two of them sustaining the other . . ."

On that very day Nicodemus, who placed the body of Jesus in his tomb, was visited by Christ "mounted on the chariot of the Cherubim."

Some time later it came to pass, according to the Apostle Peter, that Jesus commanded him to gather the rest of the apostles and all the disciples on the Mount of Olives. He did so, and they were standing on the mountain, with Jerusalem lying below them.

Then a white and shining cloud, which was like unto a flame of fire, surrounded us, and all the people of Jerusalem saw the splendour thereof and were dismayed. And we were standing in the midst of the cloud, and we saw the doors of heaven opened, and the angels of light ascending and descending upon a ladder of light. And we saw our Lord standing at the foot of the ladder, and He was wishing to ascend unto His throne of radiant glory, and He stretched out His holy right hand, and blessed us.... And again, we saw a chariot of light descend from heaven upon the wings of the Cherubim, and with it were thousands and tens of thousands of thousands of angels.

And I, Peter, saw the race of the angels, and their apparel, and their appearance, and their names. And I saw, moreover, a gathering together of the armies of the angels with flaming chariots of fire, and they were mounted upon horses of light; and when the children of men looked upon them, their eyes were carried away by the sight thereof.... And the heavens, and the earth, and the air were filled with the angels, and with the multitudes thereof. And, moreover, great numbers of the people of Jerusalem at that time saw things which I was seeing, and there were with them many of the Jews who had transgressed against righteousness, and who did not believe in our Lord Jesus Christ. And they stood at that time at the place where they could readily see these things, and fear and dismay came upon them all, and they glorified God. Then some of them who were doubtful concerning what they had seen wished to go up into the Mount of Olives, and there came upon them tongues of fire and burned up many of them.

And again, I saw a cloud in the form of a bow which appeared among the clouds, and upon it was a tabernacle of light, and in the innermost part thereof sat the holy Virgin Mary, who gave birth to our Lord in the flesh. And angels surrounded her, and in their hands were swords and spears of fire.... And again, I saw a cloud over the Mount of Olives, and I heard the voice of my Lord and God, saying thereunto, "Stand still in thy place, 0 Mount, that thou mayest be a witness for me to My ascent from thee into the heavens; and know, 0 mountain, that no one shall ascend from thee except Myself into heaven until My second coming." And after this my Lord and God took me by the hand, and raised me up from my knees, ... and He said unto me, "Awake, 0 Peter, ... be thou a witness to that which thou hast seen, and be strong to conceal everything." And again I saw the Cherubim come, and with them were chariots of light, and when I had seen the angels I was dismayed. Then our Lord ascended upon a chariot of the Cherubim, and the clouds bore Him away, and I heard the sounds of trumpets sounding.... And I Peter and my brethren the disciples watched the chariot whereon Christ our God had mounted until it entered into the first heaven, and then I saw the gates of heaven close.

And for us, we remained upon the Mount of Olives until the time of evening, and we prayed on the spot wherefrom we saw our Lord ascend into heaven, and we also prayed upon the place where the chariot rested, and upon the place whereon stood the tabernacle in which we had seen our Lord. Then we the Apostles came down from the holy mountain, and entered into the city of Jerusalem.... And every day the Apostles used to go up into the Mount of Olives at morn and at eve and pray there.[131]

In another Christian text it is said that before the Virgin Mary died she prayed, wishing that the apostles pay her a last visit. Accordingly the Twelve were transported by clouds of light to Bethlehem from different parts of the world. When the Jews learned about their presence, they requested the Roman procurator to send from Jerusalem a tribune of the soldiers against the apostles. But before the soldiers reached Bethlehem, the Holy Spirit said to the apostles: "Go forth therefore from Bethlehem, and fear not: for, behold, by a cloud I shall bring you to Jerusalem; for the power of the Father, and the Son, and the Holy Spirit is with you."

The apostles therefore rose up immediately, and went forth from the house, carrying the bed of the Lady the mother of God, and directed their course to Jerusalem; and immediately, as the Holy Spirit had said, being lifted up by a cloud, they were found in Jerusalem in the house of the Lady. And they stood up, and for five days made an unceasing singing of praise. And when the tribune came to Bethlehem, and found there neither the mother of the Lord nor the apostles, he laid hold of the Bethlehemites, saying to them: Did you not come telling the procurator and the priests all the signs and wonders that had come to pass, and how the apostles had come out of every country? Where are they, then? ... And after five days it was known to the procurator, and the priests, and all the city, that the Lord's mother was in her own house in Jerusalem, along with the apostles, from the signs and wonders that came to pass there. And a multitude of men and women and virgins came together, and cried out: Holy Virgin, that didst bring forth Christ our God, do not forget the generation of men. And when these things came to pass, the people of the Jews, with the

---

[131] Sir Ernest A. Wallis Budge, op. cit. note 66, pp. 467-481.

priests also, being the more moved with hatred, took wood and fire, and came up, wishing to burn the house where the Lord's mother was living with the apostles. And the procurator stood looking at the sight from afar off. And when the people of the Jews came to the door of the house, behold, suddenly a power of fire coming from within, by means of an angel, burnt up a great multitude of the Jews. And there was great fear throughout all the city; and they glorified God, who had been born of her....

And after all these wonderful things had come to pass through the mother of God, ... the Holy Spirit said to us: ... Cast incense, because Christ is coming with a host of angels; and, behold, Christ is at hand, sitting on a throne of cherubim. And while we were all praying, there appeared innumerable multitudes of angels, and the Lord mounted upon cherubim in great power; and, behold, a stream of light coming to the holy virgin.... And then the face of the mother of the Lord shone brighter than the light, and she rose up and blessed each of the apostles with her own hand, and all gave glory to God; and the Lord stretched forth His undefiled hands, and received her holy and blameless soul.... And Peter, and I John, and Paul, and Thomas, ran and wrapped up her precious feet for the consecration; and the twelve apostles put her precious and holy body upon a couch, and carried it.[132]

And, behold, a new miracle. There appeared above the bier a cloud exceeding great, like a great circle which is wont to appear beside the splendour of the moon; and there was in the clouds an army of angels sending forth a sweet song, and from the sound of the great sweetness the earth resounded.[133]

And, behold, while they were carrying her, a certain wellborn Hebrew, Jephonias by name, running against the body [to throw it down to the ground], put his hands upon the couch; and, behold, an angel of the Lord by invisible power, with a sword of fire, cut off his hands from his shoulders, and made them hang about the couch, lifted up in the air.[134]

And when the apostles raised the bier, part of him hung, and part of him adhered to the couch; and he was vehemently tormented with pain, while the apostles were walking and singing. And the angels who were in the clouds smote the people with blindness.[135]

And when this miracle had been done, the apostles carried the couch, and laid down her precious body in Gethsemane in a new tomb.[136]

---

[132] Alexander Walker, trans., *Translations of The Writings of the Fathers Down to A.D. 325*, in Alexander Roberts and James Donaldson, eds., *Ante-Nicene Christian Library*, Vol. 16 (Edinburgh: T. & T. Clark, 1873), pp. 504-513.
[133] Ibid., p. 527.
[134] Ibid., p. 513.
[135] Ibid., p. 528.
[136] Ibid., p. 514.

In the seventh month after the death, i.e. on 15th of Mesore, we reassembled at the tomb and spent the night watching and singing.[137]

It came to pass after that we reached the 16th of Mesore, and were gathered ... at the tomb. We saw lightnings and were afraid. There was a sweet odour and a sound of trumpets. The door of the tomb opened: there was a great light within. A chariot descended in fire: Jesus was in it; he greeted us.

He called into the tomb: Mary, my mother, arise! And we saw her in the body, as if she had never died. Jesus took her into the chariot. The angels went before them. A voice called, "Peace be to you, my brethren."[138]

And the apostles being taken up in the clouds, returned each into the place allotted for his preaching, telling the great things of God, and praising our Lord Jesus Christ, who liveth and reigneth with the Father and the Holy Spirit, in perfect unity, and in one substance of Godhead, for ever and ever. Amen.[139]

So far as the UFO controversy is concerned, in the light of the above excerpts one is inclined to agree with the American evangelist William A. (Billy) Sunday: "When the consensus of scholarship says one thing and the Word of God another, the consensus of scholarship can go plumb to hell for all I care."

Theologians and scientists—who scoff at the UFO reports—definitely can go straight down to hell in the light of the above reports. Encounters with chariots of light, with men in shining garments, and with swords and spears of fire that burned up people, cannot be "easily explained" away.

Apparently the authorities of this world are not willing to admit to the public that their military technology could offer no protection if for some reason an extraterrestrial power would decide to intervene overtly into our affairs. As the admission of the existence of a higher power would certainly create a drastic shift in loyalty, the authorities in power today, just like the authorities of the past, seem to prefer to keep the public in the dark. So they not only keep withholding vital information from the people regarding visitations from outer space, but at the same time confuse them by debunking and ridiculing such reports.

But if only one government will take seriously the information hidden in the sacred writings of the world, it will have the potential to present the rest of the world with a formidable scientific and technological surprise. Consequently the message of the scriptures that man's seed or genetic constitution is God, the creator of the universe, should be taken seriously and studied extensively by any responsible government, because on the exploitation of that arcane knowledge will depend national power and mankind's future.

---

[137] James, op. cit. note 126, p. 197.
[138] Ibid., p. 200.
[139] Walker, op. cit. note 132, p. 530.

In a speech before the American Political Science Association John Foster Dulles communicated the correct message when he said:

> Our future greatness and our power lie, not in aping the methods of Soviet communism, not in trying to contain them—and us—within walls of steel, but in demonstrating, contrastingly and startlingly, the infinitely greater worth of practices that derive from a spiritual view of the nature of man.

Pythagoras, in his Fourth Table, says: How wonderful is the agreement of Sages in the midst of difference! They all say that they have prepared the Stone out of a substance which by the vulgar is looked upon as the vilest thing on earth. Indeed, if we were to tell the vulgar herd the ordinary name of our substance, they would look upon our assertion as a daring falsehood. But if they were acquainted with its virtue and efficacy, they would not despise that which is, in reality, the most precious thing in the world. God has concealed this mystery from the foolish, the ignorant, the wicked, and the scornful, in order that they may not use it for evil purposes.

The Glory of World

Chapter Eight

SUMMARY

The preceding arguments indicate that an image analogy holds the solution to the mystery of the origin of the universe and man's place in it. While the use of the tree as an image of the universe may seem elementary, we should keep in mind that image analogy proved itself to be a potent tool in science when the fall of an apple suggested to Sir Isaac Newton the law of universal gravitation.

In the final analysis reasoning by analogy and the world views of mankind reveal an ancient cosmological model which explains the origin and the nature of the universe in terms of a perpetual seed that acted on the pre-cosmic elementary particles of chaos in order to give life, mass and form to the universe.

The main points of this seed cosmology may be summarized as follows:

• The cosmic seed, as well as the elementary particles of matter, are without origin. It is the cosmic seed, however, that acts on the formless elementary particles in order to give life to the universe.

• In its potential state the cosmic seed contains all the qualities required to the creation of the universe from the preexistent elementary particles of chaos. When it germinates, it acts on the particles of chaos, and gives life to the universe by reducing the state of chaos to order.

• The cosmic seed's progressive development into the manifest universe is completed with the creation of man.

• Because man constitutes the cosmic system's end product or output, he contains in his body a copy, image, or reproduction of that parent cosmic seed which input generated the universe for the production of man in its own image.

• Being the cause of the universe, human life is supernatural relative to the universe it generated for the production of human beings in its own image. By definition supernatural causes are outside the ability of science to investigate, but in reality it is not human life that exists outside the scope of science, but superhuman life.

• The Scientific Theory of Creation, namely that human life constitutes both the input and the output of the cosmic system—or the universe's origin and end product—is falsifiable pending the demonstration that the law of biogenesis is invalid, i.e. "The principle that a living organism can only arise from other living organisms similar to itself (i.e. that like gives rise to like) and can never originate from nonliving material." (*The Oxford Dictionary of Biology*, 4th ed., Oxford University Press, 2000.)

BIBLIOGRAPHY

Allegro, John M. *The Sacred Mushroom and the Cross*. 2nd pr. New York: Bantam Books, 1971.

Beier, Ulli. *The Origin of Life and Death*. London, Ihadan, Nairobi: Heinemann, 1969.

Besant, Annie. *The Story of the Great War*. London: Theosophical Publishing Society, 1899.

Blavatsky, Helena P. *Anthropogenesis. The Secret Doctrine*. Vol. II. Pasadena, CA: Theosophical University Press, 1970.

Blum, Harold F. *Time's Arrow and Evolution*. 3rd ed. Princeton, NJ: Princeton University Press, 1968.

Browne, Lewis. *The World's Great Scriptures*. New York: The Macmillan Company, 1946.

Budge, Sir Ernest A. Wallis, trans. *The Book of the Saints of the Ethiopian Church*. Vol. II. Cambridge: University Press, 1928.

Budge, Sir Ernest A. Wallis, trans. *The Contendings of the Apostles*. Vol. II. London: Henry Frowde, 1901.

Budge, Sir Ernest A. Wallis. *Egyptian Ideas of the Future Life*. London: Kegan Paul, Trench, Trubner & Co., Ltd., 1900.

Busenbark, Ernest. *Symbols, Sex and the Stars*. New York: Truth Seeker Co., 1949.

Cadoux, Cecil John. *The Early Church and the World*. Edinburgh: T. & T. Clark, 1925.

Chetty, D. Gopaul. *New Light on Indian Philosophy*. New York: E. P. Dutton & Co., MCMXXIII.

Conger, George Perrigo. *Theories of Macrocosm and Microcosm*. New York: Russell & Russell, 1967.

Cook, Roger. *The Tree of Life: Image for the Cosmos*. New York: Avon Books, 1974.

Cowell, E. B., ed. *The Jataka Or Stories of Buddha's Former Births*. E. B. Cowell and W. H. D. Rouse, trans. Vol. VI. Cambridge: University Press, 1907.

Crawley, Ernest. *The Mystic Rose*. London: Watts & Co., 1932.

Crick, Francis. "Seeding the Universe." *Science Digest*, November 1981, pp. 82 117.

Deane, William J. *Pseudepigrapha*. Edinburgh: T. & T. Clark, 1891.

Duchesne-Guillemin, Jacques. *Symbols and Values in Zoroastrianism*. New York: Harper & Row, 1966.

Edwardes, Allen. *Erotica Judaica*. New York: The Julian Press, 1967.

Edwardes, Allen, and Masters, R. E. L. The Cradle of Erotica. New York: The Julian Press, 1963.

Eliade, Mircea. *Patterns in Comparative Religion*. New York: Meridian Book, 1965.

Eliade, Mircea. *Yoga*. Princeton, NJ: Princeton University Press, 1969.

Forke, Alfred. *The World-Conception of the Chinese*. Arno Press, 1975.

Eppstein, John. *The Catholic Tradition of the Law of Nations*. London: Burns Oates & Washbourne Ltd., 1935.

Frankfort, Henri. *Kingship and the Gods.* Samuel Noah Kramer, pref. Chicago & London: The University of Chicago Press, 1948, 1978.

Frazer, Sir James George. *The Golden Bough*. Vol. I. New York: Macmillan Co., 1955.

Ginzberg, Louis. *The Legends of the Jews*. Henrietta Szold, trans. Philadelphia: The Jewish Publication Society of America, 1909, 1913.

Grimal, Pierre, ed. *Larousse World Mythology*. 2nd ed. London: Paul Hamlyn, 1969.

Hahm, David E. *The Origins of Stoic Cosmology*. Ohio State University Press, 1977.

Hall, Manly P. *Man: The Grand Symbol of the Mysteries*. Los Angeles: The Philosophical Research Society, Inc., 1972.

Hall, Manly P. *The Secret Teachings of All Ages*. 19th ed. Los Angeles: The Philosophical Research Society, Inc., 1973.

Hennecke, Edgar, and Schneemelcher, Wilhelm, eds. *New Testament Apocrypha*. Vol. 1. R. McL. Wilson, trans. Philadelphia: The Westminster Press, 1963.

Hunt, H. A. K. *A Physical Interpretation of the Universe: The Doctrines of Zeno the Stoic*. Melbourne University Press, 1976.

Ishihara, Akira, and Levy, Howard S. *The Tao of Sex*. Har/Row Books, 1970.

Jacobi, Jolande, ed. *Paracelsus: Selected Writings*. Norbert Guterman, trans. Princeton, NJ: Princeton University Press, 1969.

James, Montague Rhodes, trans. *The Apocryphal New Testament*. Oxford: Clarendon Press, 1924, 1960.

Jones, Rufus M. *Studies in Mystical Religion*. London, 1909.

Jung, Carl G. *Aion*. H. F. C. Hull, trans. 2nd ed. Princeton, NJ: Princeton University Press, 1970.

Jung, Carl G. *Alchemical Studies*. R. F. C. Hull, trans. 2nd ed. Princeton, NJ: Princeton University Press, 1970.

Jung, Carl G. *Mysterium Coniunctionis*. R. F. C. Hull, trans. 2nd ed. Princeton, NJ: Princeton University Press, 1970.

Jung, Carl G. *Psychology and Alchemy*. H. F. C. Hull, trans. 2nd ed. Princeton, NJ: Princeton University Press, 1970.

King, Francis. *Sexuality, Magic and Perversion*. Secaucus, NJ: Citadel Press, 1974.

Leslie, Desmond, and Adamski, George. *Flying Saucers Have Landed*. New York: The British Book Centre, January 1953.

Lockyer, Herbert. *All the Parables of the Bible*. Grand Rapids, MI: Zondervan Publishing House, 1963.

Mann, Felix. *Acupuncture*. New York: Vintage Books, 1973.

Mumford, Jonn. *Sexual Occultism*. Saint Paul, MN: Llewellyn Publications, 1975.

Nicolai, G. F. *The Biology of war*. New York: The Century Co., 1918.

Nikhilananda, Swami. *The Upanishads*. New York: Harper & Row, 1964.

Ossendowski, Ferdinand. *Beasts, Men and Gods*. E. P. Dutton & Company, 1922.

Perry, Whitall N. *A Treasury of Traditional Wisdom*. New York: Simon and Schuster, 1971.

Philostratus, Flavius. *Life and Times of Apollonius of Tyana*. Charles P. Eells, trans. Stanford University, CA: Stanford University Press, 1923.

Polski, G. "Are we alone in the Universe?" *Sovietskaya Rossiya*. Trans. and repr. in Moscow News. No. 47, 1978.

Prabhavananda, Swami, and Isherwood, Christopher, trans. *The Song of God: Bhagavad-Gita*. 3rd ed. New York: Mentor Book, 1955.

Prabhavananda, Swami, and Manchester, Frederick, selected and trans. *The Upanishads: Breath of the Eternal*. New York: Mentor Book, 1957.

Ravenscroft, Trevor. *The Spear of Destiny*. New York: Bantam Books, 1974.

Rawson, Philip. *Tantra: The Indian Cult of Ecstasy*. New York: Avon Books, 1973.

Regardie, Israel. *The Philosopher's Stone*. 2nd ed. Saint Paul, MN: Llewellyn, 1970.

Reichel-Dolmatoff, Gerardo. *Amazonian Cosmos: The Sexual and Religious Symbolism of the Tukano Indians*. Chicago University Press, 1971.

Roheim, Geza. *Australian Totemism: A Psycho-Analytic Study in Anthropology*. M. D. Eder, intr. London: G. Allen & Unwin Ltd., 1925.

Seldes, George, comp. *The Great Quotations*. Secaucus, NJ: Castle Books, 1977.

Van Den Broek, R. *The Myth of the Phoenix*. Leiden: E. J. Brill, 1972.

Vawter, Bruce. *On Genesis: A New Reading*. Garden City, NY: Doubleday, 1977.

Von Franz, Marie-Louise, ed. *Aurora Consurgens: Commentary*. R. F. C. Hull and A. S. B. Glover, trans. New York: Pantheon Books, 1966.

Walker, Alexander, trans. *Translations of The Writings of the Fathers Down to A.D. 325*. Alexander Roberts and James Donaldson, eds. Ante-Nicene Christian Library. Vol. 16. Edinburgh: T. & T. Clark, 1873.

Walker, Benjamin. *Hindu World: An Encyclopedic Survey of Hinduism*. Vol. I. London: George Allen & Unwin Ltd., 1968.

Wilhelm, Richard, trans. and expl. *The Secret of the Golden Flower: A Chinese Book of Life*. Carl G. Jung, forew. and comm. New York: Harvest Book, 1962.

Zaehner, R. C. *The Dawn and Twilight of Zoroastrianism*. New York: G. P. Putnam's Sons, 1961.

Zeller, Eduard. *Outlines of the History of Greek Philosophy*. 13th ed. New York: Humanities Press, 1951.

Zimmer, Heinrich. *Myths and Symbols in Indian Art and Civilization*. Joseph Campbell, ed. New York: Harper & Row, 1962.

Made in the USA
Las Vegas, NV
14 May 2023